New Challenges in Residual Stress Measurements and Evaluation

Edited by Caterina Casavola, Claudia Barile, Vincenzo Moramarco and Giovanni Pappalettera

Published in London, United Kingdom

IntechOpen

Supporting open minds since 2005

New Challenges in Residual Stress Measurements and Evaluation
http://dx.doi.org/10.5772/intechopen.78830
Edited by Caterina Casavola, Claudia Barile, Vincenzo Moramarco and Giovanni Pappalettera

Contributors
Jiangchao Wang, Qingya Zhang, Hong Zhou, Byung-Ik Kang, Frank Liou, Heng Liu, Sanichiro Yoshida, Tomohiro Sasaki, Madalina Rabung, Michael Becker, Meisam Amiri, Melanie Kopp, Ralf Tschuncky, Ines Veile, Klaus Szielasko, Fabian Weber, Miriam Weikert-Müller, Lorenzo Bertelli, Alessio Benincasa, Emilio Valentini, Simone Gulisano, Caterina Casavola, Claudia Barile, Vincenzo Moramarco, Giovanni Pappalettera

First published in London, United Kingdom, 2020 by IntechOpen
IntechOpen is the global imprint of INTECHOPEN LIMITED, registered in England and Wales, registration number: 11086078, 7th floor, 10 Lower Thames Street, London, EC3R 6AF, United Kingdom
Printed in Croatia

British Library Cataloguing-in-Publication Data
A catalogue record for this book is available from the British Library

Additional hard and PDF copies can be obtained from orders@intechopen.com

New Challenges in Residual Stress Measurements and Evaluation
Edited by Caterina Casavola, Claudia Barile, Vincenzo Moramarco and Giovanni Pappalettera
p. cm.
Print ISBN 978-1-78984-951-6
Online ISBN 978-1-78984-952-3
eBook (PDF) ISBN 978-1-83880-280-6

For EU product safety concerns:
IN TECH d.o.o., Prolaz Marije Krucifikse Kozulić 3, 51000 Rijeka, Croatia, info@intechopen.com or visit our website at intechopen.com.

Meet the editors

Prof. Casavola is a full professor of Mechanical Design at Politecnico di Bari and she teaches *Design of Innovative Materials*, *Experimentation on Aerostructures and Experimental Mechanics*. Prof Casavola's research activity is concerned with mechanical characterization of innovative materials, experimental stress analysis on real components (1:1 scale) and *in situ* inspection, study of residual stress generation, evolution and measurements (hole drilling method both with strain gage rosette and ESPI, X-Ray diffractometer, Barkhausen noise), and monitoring of electronic board and components by means of speckle interferometry. Prof. Casavola is the Scientific Director of the Laboratory of Static and Dynamic tests, Laboratory of Residual Stresses, Laboratory of Experimental Stress Analysis, Laboratory EMILIA (Experimental Mechanics Integrated Laboratory In Aerospace), and the Laboratory of Telemetry at Politecnico di Bari.

Claudia Barile has been a Lecturer in the Department of Mechanics, Mathematics and Management at the Politecnico di Bari since April 2017. She achieved her Master's degree in Mechanical Engineering Magna cum laude in July 2008 at the Politecnico di Bari. She received her PhD title on February 2012 in Advanced Production Systems at the Politecnico di Bari. Her research activities are mainly focused on the mechanical characterization of materials (composites, metals, polymers) with both traditional and innovative experimental techniques.

Dr. Vincenzo Moramarco has been a Lecturer at the Department of Mechanics, Mathematics and Management at the Politecnico di Bari since April 2017. In 2010 he received the PhD title (Doctor Europaeus) in Bioengineering and Biomechanical Engineering from Scuola Interpolitecnica di Dottorato (Turin). His research activities are mainly focused on the experimental mechanical characterization and numerical modelling of materials both under static and dynamic load conditions. His work deals with various families of materials (e.g. polymers, metals, foams, biological hyper-elastic materials, sandwich panels and materials obtained by additive manufacturing techniques). He is familiar with several experimental techniques (e.g. thermography, optical methods, X-ray diffractometry) to measure stress, deformation and residual stresses of materials and components.

Giovanni Pappalettera has been a Lecturer in the Department of Mechanics, Mathematics and Management at the Politecnico di Bari since 2015. He achieved his Master's degree in Physics Magna cum laude in 2003 at the Università di Bari. He received his PhD title on 2011 in Mechanical Design at the Scuola Interpolitecnica di Dottorato in Torino. His research activities are mainly focused on optical methods for mechanical characterization of materials and residual stress measurements, 3D contouring and acoustic emission analysis in fracture mechanics. He is co-inventor of three patented devices.

Contents

Preface

Residual stress experimental analysis is a topic of great relevance where a lot of effort over the years has led to the development of a considerable number of experimental methods that can be applied conveniently to several different measurement situations. In spite of the large number of papers that have been published in the last decades, the subject is far from be considered as concluded, on the contrary, a lot of research is currently being undertaken in many different directions such as the improvement of the resolution of the measurement, the capability of measuring residual stress even in the core of the component as well as increasing the level of portability and usability of the instrumentation. At the same time, numerical methods have also been under study in order to develop powerful and accurate analytical instruments to predict the level of introduced residual stress along each step of the complex production chain starting with material production and ending with the final manufactured part. The increasing number of materials for which residual stress evaluation is required (nowadays also including biomaterials, additive manufactured materials and composites) makes this challenge even wider. Far from being a comprehensive review of the research performed on this topic, this book specifically aims to provide updated information on some of the most recent advancements in this field. The book opens with an introductory chapter that provides the reader with an overview of the most recent progresses by identifying the most appealing routes for future developments. Some of that will then be discussed in the following chapters of this book divided into two sections. In the first section research advances in experimental methods for residual stress measurements are discussed while the second section is devoted to the advances in residual stress modelling and prediction. In Chapter 1, a very innovative and promising approach is illustrated based upon the synergic use of the Electronic Speckle Pattern Interferometry and the acoustic method. The evaluation of the sound velocity is proposed as a tool for evaluating the elastic modulus and, consequently, the local residual stress while the full-field character of ESPI is exploited to obtain a 2D distribution of the stress field. In Chapter 2, two alternative non-destructive approaches are compared based on the analysis of the magnetic and acoustic response of a material as it is modified by the presence of locked-in residual stress. In particular, interesting observations can be found in relationship to the possibility to evaluate RS at the macro- and micro-scale.

In Chapter 3, experimental analysis using the classical approach of the hole-drilling methods is presented. Even being a well-known, consolidated and standardized method, the authors present a novel approach that is promising in overcoming some current limitations and improving the level of attainable accuracy. In Chapter 4, a coupled thermo-mechanical finite element model is presented allowing determination of residual stress distribution and distortion in a stainless steel part manufactured by direct metal deposition, with an experimental validation presented as well. In Chapter 5, an Inverse Finite Element Method is proposed for evaluating residual stress in thick butt weld joints as well as a validation procedure based on the contour method. Finally, in Chapter 6, a study of residual stress simulation is presented for the case of a steel cylindric block, once calculated the simulated stress field is used as an input for accurate durability assessment.

As editor I would like to express my gratitude to all the authors that have contributed their work and their experience to this book. I also would like to acknowledge my co-editors Claudia Barile, Vincenzo Moramarco and Giovanni Pappalettera for their valuable contribution. Our greatest wish is that this publication is useful for experts in the residual stress field and inspiring for young researchers approaching this challenging field.

Caterina Casavola, Claudia Barile, Vincenzo Moramarco and Giovanni Pappalettera
Department of Mechanical,
Mathematics and Management,
Politecnico di Bari,
Bari, Italy

Section 1

Introduction

Chapter 1

Introductory Chapter: New Challenges in Residual Stress Measurements and Evaluation

Caterina Casavola, Claudia Barile, Vincenzo Moramarco and Giovanni Pappalettera

1. Definition of residual stresses and their relevance

Residual stresses (RS) are stresses present inside the materials and the structures even in absence of any applied external load. Residual stresses arise as a consequence of the complex sequence of steps starting from the material production, through machining, to final joining of subparts allowing manufacturing the final component [1]. Each production stage, in fact, such as casting, forming, joining, cutting, grinding, polishing, welding, heat treating, and so on, contributes to modify the pre-existing state of stress, and all of them add their specific contribution to determine the final residual stress field in the manufactured part [2].

Knowledge, control, and modification of the level of RS inside a given material or part become of capital importance by taking into consideration that residual stresses can impact greatly on the mechanical strength of the material, on its dimensional correspondence to design specifications, as well as on the corrosion resistance and on the fatigue life and durability of the part. It is well-known, for example, that tensile RS play a relevant role in reducing the fatigue life of the component so that several specific stress relaxation treatments have been developed in order to mitigate or eliminate their presence [3]. On the other side, the presence of a compressive stress field can have positive effects, and, due to this reason, some processes such as shot peening or laser shock peening have been developed to allow their introduction.

With respect to the recent past, it can be observed that mechanical designers have, nowadays, a greater awareness about the necessity to take into account also the presence of residual stress in their project. Relevance of RS, in fact, is now well assessed not only at academic level but also on the industrial side. At the same time, the multiplication and the evolution of the production techniques, the appearance of new disruptive manufacturing processes such as the additive manufacturing, the availability of more and more performing surface treatment techniques, the introduction of new coating technologies, and the request for increasingly better performing components result in a demand for more updated methods for residual stress evaluation.

The main aim of this book is to provide the reader with an overview of the principal challenges that research in terms of residual stress evaluation is currently facing including current limitations and potential future developments. At the same time, the reader will have an overview of the different materials, production technologies, and fields of applications where interest in RS evaluation

IntechOpen

is of capital importance but at the same time still not conveniently fulfilled by traditional and consolidated approaches.

2. Current trends in residual stress experimental analysis and numerical evaluation

The importance, and at the same time, the complexity, of obtaining an accurate experimental evaluation of the residual stress inside a material has led, along the years, to the development of many different approaches. Some of them such as the hole drilling method (HDM) and X-ray diffraction (XRD) are, nowadays, mature and widespread techniques ruled by specific standards [4, 5]. Even if they can be considered as benchmark techniques, it should be taken into account that, in many situations, they have some inherent limitations that cannot be discarded in some specific applications.

Hole drilling method, for example, cannot be conveniently used to determine the presence of residual stress too much far from the drilled surface. X-ray diffractometer is limited to the determination of superficial residual stress. Both the approaches cannot guarantee high surface resolution of the residual stress measurement.

More in general, it should be understood that each method developed for the residual stress evaluation has its specific advantages and drawbacks that make a specific approach appealing for a given application problem and inapplicable in some other situations. Having this in mind, however, it should be clear that current efforts are spent in the direction to increase the resolution of the measurement, both on the surface and in-depth, and to extend the capability to determine the residual stress present in the core of a given component. At the same time, also the development and improvement of nondestructive techniques for residual stress determination become of capital importance because this would expand the possibility to perform in situ evaluation.

Additionally, a further challenge is in the direction of adopting less expensive equipment and to reduce, at the same time, the cost and the time required for a single measurement [6]. A final direction of evolution of the research in the residual stress field is toward the RS characterization at the micro- and nanoscale. This is of interest by taking into account that, when dealing with high cycle fatigue and very high cycle fatigue conditions, the damage accumulation occurs at the grain level. Moreover, the introduction of micro- and nanomechanical systems requires an enhancement of the capabilities of characterization over this range scale. In the direction of improving the on-surface resolution, it worth mentioning the great effort that was done in the last years in the development of the contour method [7] that allows obtaining a complete map distribution of the residual stress along the cut plane, but it is limited to materials that can be cut by EDM machine, and it exhibits a higher level of uncertainty close to the edges of the measured surfaces. Magnetic methods such as magnetic Barkhausen noise [8] and acoustic Barkhausen noise [9] can be adopted as well to obtain residual stress map on a component, and current efforts are spent in the direction of accurate calibration procedures that take into account other factors such as the hardness, the crystallographic nature, or the complexity of the stress field that can impact on the measurement. Nevertheless, their adoption is limited to magnetic materials. The problem of increasing the depth of evaluation of residual stresses can be conveniently afforded, for example, by implementing synchrotron X-ray diffraction or neutron diffraction [10]; in both cases, however, very expensive equipment is required as well as long-term planning of the experimental campaign to access to the dedicated facilities. As an alternative some modifications of the hole drilling method such as the deep hole drilling method [11] and the slitting method [12] are studied and under development for in-depth measurements. In the direction of developing new nondestructive approaches, it appears now promising the possibility to combine optical and acoustic methods [13] to evaluate residual stress through the detection of the variation of the elasticity constant of the material. This is, incidentally, along another path of development followed by scholars in the last years aiming to combine multiple approaches in a single hybrid method (e.g., hole drilling/ring core). Capability to measure residual stress at the micro- and nanoscale is showing promising results, based, for example, on the adoption of the focused ion beam (FIB)-DIC micro-ring-core technique. This approach is very promising to be adopted for the evaluation of the type I, II, and III residual stresses. Raman spectroscopy, as well, is attracting a lot of interest in view of its capability to obtain microscale residual stress measurements [14]. It is worth mentioning, additionally, that efforts in improvement in experimental capability of residual stress evaluation impact also on the development and more accurate validation of modeling for numerical RS prediction. In this area the major challenges can be identified in the direction to extend the number of manufacturing processes and post-processes that can be modeled as well as to make the evaluation process faster and more accurate by introducing, for example, effects and features such as the kinematic hardening [15], phase transformation, and anisotropy.

Author details

Caterina Casavola*, Claudia Barile, Vincenzo Moramarco and Giovanni Pappalettera
Dipartimento di Meccanica, Matematica e Management, Politecnico di Bari, Bari, Italy

*Address all correspondence to: caterina.casavola@poliba.it

References

[1] Casavola C, Pappalettera G, Pappalettere C, Tursi F. Effects of strain's error on residual stresses calculated by HDM. Experimental and Applied Mechanics. 2013;**4**:395-402

[2] Totten G, Howes M, Inoue T. Handbook of Residual Stress and Deformation of Steel. Materials Park, Ohio (US): ASM International; 2002

[3] Withers PJ. Residual stress and its role in failure. Reports on Progress in Physics. 2007;**70**(12):2211-2264

[4] ASTM. E837-08 Standard Test Method for Determining Residual Stresses by the Hole-Drilling Strain Gage Methods. West Conshohocken (US): American Society for Testing and Materials; 2008

[5] EN15305:2008 Non-Destructive Test Method for Residual Stress Analysis by X Ray Diffraction. Bruxelles: CEN European Committee for Standardization; 2008

[6] Barile C, Casavola C, Pappalettera G, Pappalettere C. Analysis of the effects of process parameters in residual stress measurements on titanium plates by HDM/ESPI. Measurement. 2014;**48**:220-227

[7] Prime MB. Cross-sectional mapping of residual stresses by measuring the surface contour after a cut. Journal of Engineering Materials and Technology. 2001;**123**(2):162-168

[8] Casavola C, Pappalettere C, Tursi F. Calibration of Barkhausen noise for residual stress measurement. Experimental and Applied Mechanics. 2012;**4**:255-266

[9] Buttle DH, Dalzell W, Scruby CB, Langman RA. Comparison of three magnetic techniques for biaxial stress measurement. In: Proceedings of Review of Progress in Quantitative Nondestructive Evaluation. Brunswick (US): Plenum Publishing; 1989

[10] ISO/TS 21432. Non-destructive testing-standard Test Method for Determining Residual Stresses by Neutron Diffraction. ISO/TS Technical Specification; 2005

[11] DeWald AT, Hill MR. Improved data reduction for the deep-hole method of residual stress measurement. Journal of Strain Analysis. 2003;**38**(1):65-78

[12] Prime MB. Residual stress measurement by successive extension of a slot: The crack compliance method. Applied Mechanics Reviews. 1999;**52**(2):75

[13] Yoshida S, Sasaki T, Usui M, Sakamoto S. Residual stress analysis based on acoustic and optical methods. Materials. 2016;**9**(2):122

[14] Liu D, Flewitt PEJ. Raman measurement of stress in films and coatings. In: Yarwood J, Douthwaite R, editors. Spectroscopic Properties of Inorganic and Organometallic Compounds. 1st ed. Cambridge (UK): Royal Society of Chemistry; 2014. pp. 141-177

[15] Chakherlou TN, Yaghoobi A. Numerical simulation of residual stress relaxation around a cold-expanded fastener hole under longitudinal cyclic loading using different kinematic hardening models. Fatigue & Fracture of Engineering Materials & Structures. 2010;**33**(11):740-751

Section 2

Advances in Experimental Methods for Residual Stress Measurements

Chapter 2

Opto-Acoustic Technique for Residual Stress Analysis

Sanichiro Yoshida and Tomohiro Sasaki

Abstract

Residual stress analysis based on co-application of acoustic and optical techniques is discussed. Residual stress analysis is a long-standing and challenging problem in many fields of engineering. The fundamental complexity of the problem lies in the fact that a residual stress is locked into the material and therefore hidden inside the specimen. Thus, direct measurement of residual stress in a completely nondestructive fashion is especially difficult. One possible solution is to estimate residual stress from the change in the elastic constant of the material. Residual stress alters the interatomic distance significantly large that the elastic constant is considerably different from the nominal value. From the change in the elastic constant and knowledge of the interatomic potential, it is possible to estimate the residual stress. This acoustic technique (acoustoelasticity) evaluates the elastic modulus of the specimen via acoustic velocity measurement. It is capable of determining the elastic modulus absolutely, but it is a single-point measurement. The optical technique (electronic speckle pattern interferometry, ESPI) yields full-field, two-dimensional strain maps, but it requires an external load to the specimen. Co-application of the two techniques compensates each other's shortfalls.

Keywords: acoustoelasticity, electronic speckle pattern interferometry, scanning acoustic microscopy, finite element modeling of residual stress, nondestructive residual stress analysis, heat-induced residual stress

1. Introduction

It is widely known that residual stresses are created by almost every material process and harmful to a variety of structures and devices [1]. Yet, the problem is far from being solved. Although a number of techniques have been developed to evaluate residual stresses, destructively [2–6] or nondestructively [7–20], have been developed to evaluate residual stresses, there is no single method applicable to general cases for accurate evaluations. The fundamental complexity of the problem lies in the fact that residual stresses are locked into the material and therefore hidden from observance. Unless the locking mechanism is removed, the residual stress is not visible from the outside. Under such a situation, it is especially difficult to diagnose residual stresses nondestructively. Techniques classified as nondestructive methods normally use diffractometry [9, 10, 19] or acoustic probing [7, 8, 12–16]. The techniques classified as diffractometry detect the atomic rearrangement due to residual stresses from a shift in the diffraction angle using X-ray, neutron, synchrotron, or similar radiation. The techniques classified as acoustic probing detect residual stresses from the change in the acoustic impedance due to stress-induced alternation of the elastic constant. Both are well-established methods but have limitations. Diffractometry is applicable to detection of atomic rearrangement within the penetration depth of the radiation used. Neutron and synchrotron radiations can have reasonable penetration depths of tens of mm, but a large facility is required, and access is most often an issue. X-ray diffraction instruments are relatively more accessible, but the typical penetration depth is in the range of tens of μm [19]. The acoustic waves can penetrate much further, but the coupling of the acoustic signal from the emitter to the specimen is sensitive. According to our experience, the conditions of the coupling medium (usually distilled water) such as the layer thickness and the total amount can affect the measurement. Slight shifts in the contact locations seem to cause a considerable change in the measurement as well. It is not clear if such changes are due to errors associated with the contact or actual spatial variation of the residual stress. In either case, the measurement is essentially pointwise, and therefore these methods are time-consuming if applied to a certain area of the object. These issues are especially significant for analysis of welding-induced residual stress because by nature compressive and tensile residual stresses can alternate with high spatial frequency.

Given the above situation, our approach is to employ multiple methods. We are especially interested in combining an optical technique capable of full-field analysis and an acoustic technique. It should be noted that the optical methods, including the diffractometry discussed above, measure a change in displacement (strain), whereas the acoustic methods detect a change in elastic constant. With an appropriate combination of optical and acoustic techniques, we can essentially obtain both elastic constant and strain information. In principle, through the knowledge of strain with the elastic constant, we can estimate the stress. This is contrastive to the application of diffractometry by itself. In that case, even if the method provides us with accurate strain data, the residual stress heavily depends on the elastic constant to be used. The use of the nominal value (the residual stress free value) to estimate the residual stress from the measured residual strain is somewhat contradictory. The aim of this article is to discuss the application of an optical technique known as the electronic speckle pattern interferometry and acoustic methods for assessment of residual stress induced by welding. After briefly reviewing several techniques widely used for residual stress analysis, the results of our research are presented and discussed. The presented combined method is still in the process of development. It is our intention to introduce the techniques to the readers hoping that the information is useful to them. Future directions of the research are discussed as well.

2. Techniques to evaluate residual stresses

2.1 X-ray diffractometry (XRD)

In this section the principle of X-ray diffractometry (XRD) for residual stress analysis is described briefly. A more detailed description about the technique can be found elsewhere [21]. **Figure 1a** illustrates that Bragg's law relates the angle of diffraction θ and the lattice distance as follows:

$$n\lambda = 2d\sin\theta \tag{1}$$

Here n is an integer and λ is the wavelength of the X-ray. Eq. (1) represents the condition of constructive interference that takes place when the path difference between the reflections from the top lattice plane and the second lattice plane is an

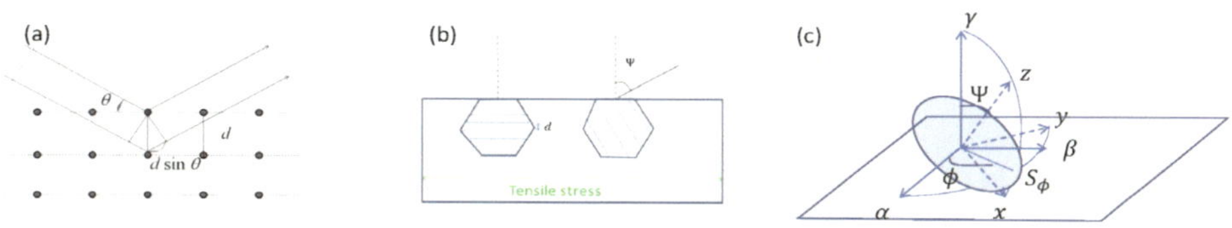

Figure 1.
(a) Bragg diffraction. (b) Change in crystallographic plane orientation and spaces due to residual stress. (c) Coordinate system αβγ affixed to specimen surface where the γ-axis is normal to specimen surface and α and β are axes of principal stress. Coordinate system xyz is affixed to grain where the z-axis is normal to lattice plane. The z-axis is rotated around the y-axis, which is in αβ-plane. S_ϕ denotes direction of in-plane strain ε_ϕ, which makes angle ϕ from the α-axis in αβ-plane.

integer multiple of λ. Under this condition, the detector (not shown in this figure) records a maximum when it is oriented at the angle of diffraction.

The XRD for residual stress analysis exploits the fact that an in-plane stress alters the distance between the neighboring lattice planes via Poisson's effect and thereby shifts the diffraction angle from the nominal (unstressed) value. **Figure 1b** illustrates the situation where an in-plane tensile stress increases the lattice distance in a grain whose lattice plane is oriented at angle Ψ to a line normal to the surface. The resultant change in the lattice distance $\Delta d_\Psi = d_\Psi - d_0$ from the initial distance d_0 causes the strain.

$$(\varepsilon_\Psi)_{zz} = \frac{d_\Psi - d_0}{d_0} \tag{2}$$

Here $(\varepsilon_\Psi)_{zz}$ is the normal strain along the z-axis in the coordinate system affixed to the grain, as **Figure 1c** illustrates. In this coordinate system, the lattice plane is parallel to the xy-plane and perpendicular to the z-axis. The $\alpha\beta\gamma$ coordinate system is affixed to the specimen where the γ-axis is normal to the specimen surface and α- and β-axes are the principal axes. The local stress has angle ϕ to the α-axis. Through the proper coordinate transformation, we can express this strain using the in-plane strain parallel to the specimen surface as.

$$\begin{aligned}(\varepsilon_\Psi)_{zz} &= \left(\varepsilon_{\alpha\alpha}\cos^2\phi + \varepsilon_{\alpha\beta}\sin 2\phi + \varepsilon_{\beta\beta}\sin^2\phi\right)\sin^2\Psi \\ &\quad + \varepsilon_{\gamma\gamma}\cos^2\Psi + \varepsilon_{\gamma\alpha}\cos\phi\sin 2\Psi + \varepsilon_{\beta\gamma}\sin\phi\sin 2\Psi.\end{aligned} \tag{3}$$

The first part on the right-hand side of Eq. (3) enclosed by the parenthesis is the in-plane strain in the direction of S_ϕ (ε_ϕ in **Figure 1**). Eq. (3) indicates that ε_ϕ is given as the slope of ε_{zz}- $\sin^2\Psi$ plot. Eq. (2) tells us the deformed lattice distance d_Ψ is proportional to ε_{zz}. These altogether indicate that the in-plane strain ε_ϕ can be given by the following differential:

$$\varepsilon_\phi = \frac{\partial(\varepsilon_\Psi)_{zz}}{\partial(\sin^2\Psi)} = \left(\varepsilon_{\alpha\alpha}\cos^2\phi + \varepsilon_{\alpha\beta}\sin 2\phi + \varepsilon_{\beta\beta}\sin^2\phi\right) \propto \frac{\partial d_\Psi}{\partial(\sin^2\Psi)}. \tag{4}$$

Since the in-plane strain ε_ϕ is connected with the in-plane stress σ_ϕ with an elastic constant, Eq. (4) indicates that σ_ϕ can be evaluated as the slope of d_Ψ vs. $\sin^2\Psi$ plot. In the present context, σ_ϕ is the in-plane residual stress of interest. Experimentally, we can evaluate σ_ϕ by changing angle Ψ and plotting the corresponding d_Ψ as a function of $\sin^2\Psi$. The deformed lattice distance d_Ψ can be determined from the corresponding shift in the diffraction angle as discussed above. When the angle Ψ is varied for this plot, the slope (4) remains unchanged.

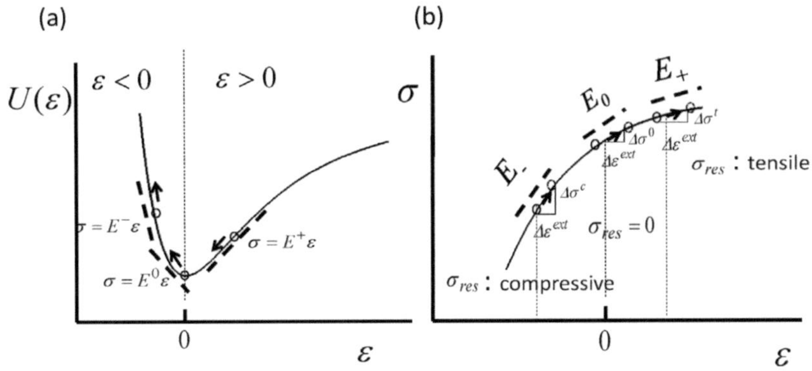

Figure 2.
(a) Interatomic potential energy curve. (b) Slope of potential energy curve.

So, d_Ψ- $\sin^2\Psi$ graph is a linear plot. Since the twice of the diffraction angle 2θ is the quantity directly measured in this type of experiment, often $2\theta - \sin^2\Psi$ graph is used to evaluate σ_ϕ. The constant of proportionality K is called the stress constant:

$$\sigma_\phi = KM \tag{5}$$

Here σ_ϕ is the residual stress, M is the slope of the $2\theta - \sin^2\Psi$ plot. K depends on the wavelength (i.e., the X-ray source line) and the lattice plane used for diffraction. As an example, for aluminum alloy 5083 with the use of Cr-Kα line for the X-ray source and the aluminum's [4 2 2] lattice plane for diffraction, $K = -168.80$ MPa/°.

2.2 Acoustoelasticity

Acoustoelasticity [14–17] evaluates residual stresses based on a change in the acoustic velocity from the nominal value. Residual stresses cause the strain so large that the elastic coefficient is altered from the nominal value. **Figure 2** illustrates the situation schematically. The strain energy curve is steeper on the short-range side of the equilibrium position (where the strain is null) than the long-range side. Consequently, the region of tensile residual stress makes the acoustic velocity lower than the nominal value, and the region of compressive residual stress makes the acoustic velocity higher. Acoustic velocity is proportional to the ratio of the elastic modulus to the density. Thus, through measurement of acoustic velocity at each point of the specimen and scanning through the entire specimen, it is possible to map out the residual stress distribution. Typically, a contact acoustic transducer is used for acoustic velocity measurement.

For quantitative analyses, the lowest order of the nonlinear terms in the elastic modulus is used. As **Figure 2** indicates, the strain energy curve is quadratic around the equilibrium (the bottom of the well). Being the first-order spatial derivative of the energy, the stress is proportional to the strain; hence, the elastic coefficient (the stress divided by the strain) is a constant. When a residual stress shifts the strain from the equilibrium, the strain energy curve is no more quadratic at that point. Hence, the elastic coefficient E becomes a function of strain and can be expressed as a polynomial expansion around the nominal value (E_0):

$$E = E_0 + C^{(3)}\varepsilon + \frac{1}{2}C^{(4)}\varepsilon^2 + \cdots + \frac{1}{(n-2)!}C^{(n)}\varepsilon^{(n-2)} + \cdots \tag{6}$$

where $C^{(n)}$ is the nth-order coefficient of the strain energy. (The exponent used for C indicates the order of the strain energy. Since the elastic constant is the slope of the strain energy, i.e., its differential, n appears on the $n-1^{th}$ term in this expression. It appears on the nth term in the potential energy expression.) Under this condition, the acoustic velocity can be expressed as follows:

$$v^{ac} = \sqrt{\frac{E_0 + C^{(3)}\varepsilon}{\rho}} \tag{7}$$

Here ρ is the density, and up to the third order of the higher-order terms in Eq. (6) is considered. The coefficient $C^{(3)}$ is called the third-order elastic constant (TOEC). When a residual stress causes the nonlinearity, the relative acoustic velocity can be expressed with the residual strain ε_{res} and the TOEC as follows:

$$v_{rel}^{ac} = \frac{v_{res}^{ac}}{v_0^{ac}} = \sqrt{\frac{E_0 + C^{(3)}\varepsilon_{res}}{E_0}} = \sqrt{1 + \frac{C^{(3)}\varepsilon_{res}}{E_0}} \simeq 1 + \frac{1}{2}\frac{C^{(3)}\varepsilon_{res}}{E_0} \tag{8}$$

Here v_{res}^{ac} and v_0^{ac} are the acoustic velocity in a residually stressed specimen and an unstressed specimen of the same material, respectively. By measuring these velocities and knowing the value of the TOEC, we can evaluate the residual strain by solving Eq. (8) for ε_{res} as.

$$\varepsilon_{res} = \frac{2E_0}{C^{(3)}}\left(v_{rel}^{ac} - 1\right) \tag{9}$$

Once ε_{res} is found, the corresponding residual stress can be evaluated with the use of the nonlinear elastic modulus expression (6).

2.3 Contact acoustic transducer and scanning acoustic microscope (SAM)

A contact acoustic transducer and scanning acoustic microscope are typical devices used for the acoustoelasticity measurement.

Figure 3a illustrates a typical contact acoustic transducer arrangement. The transducer placed on the specimen surface through a coupling medium (typically distilled water) sends a pulsed longitudinal or shear acoustic wave. The signal goes

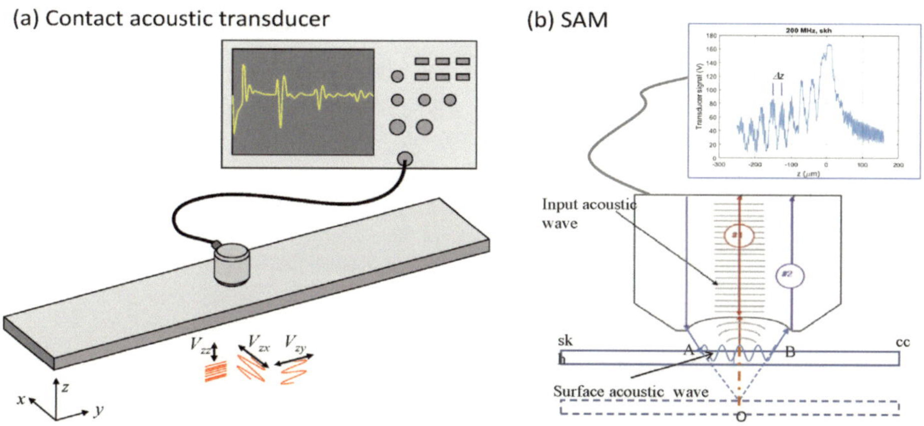

Figure 3.
(a) Typical contact acoustic transducer arrangement. (b) Working principle of SAM.

through the specimen, is reflected at the bottom surface, and returns to the transducer. The received signal shows two peaks as the insert in **Figure 3a** illustrates. The first peak represents the input signal, and the second represents the returning signal. From the time of flight, the acoustic velocity inside the specimen can be evaluated for each polarization (longitudinal or shear polarization) of the acoustic signal. Since the acoustic velocity is proportional to the square root of the elastic constant for the given polarization, the elastic constant can be estimated.

Figure 3b illustrates a typical SAM setup. Details of its operation principle can be found in a number of references [22, 23]. In short, it works as follows. The acoustic lens sends the incident acoustic wave to the specimen at an angle higher than the critical angle. Consider the two acoustic paths labeled #1 and #2 in the figure. The former represents the acoustic wave incident to the specimen surface and specularly reflected off the surface. The latter represents the acoustic reradiation at the liquid–solid interface due to the surface acoustic wave generated by the incident wave. These two acoustic waves interfere with each other. The acoustic lens is then moved toward the specimen. As this happens, the voltage signal from the transducer (not shown in the figure) placed on top of the lens undergoes a series of crests and troughs (as the path difference goes through constructive and destructive interference). This oscillatory voltage pattern is called the V(z) curve [22]. The insert in **Figure 3b** is a sample V(z) curve. Since the frequency is fixed at the acoustic source, the acoustic path length (the path length in the unit of the wavelength) over AB depends on the phase velocity of the surface wave. Thus, the interval of these peaks (Δz) is related to the velocity of the surface acoustic wave relative to the acoustic velocity in the coupling water as follows.

$$V_s = \frac{V_w}{\sqrt{1 - \left(1 - \frac{V_w}{2\Delta z} \cdot f\right)^2}} \tag{10}$$

where V_s is the surface acoustic wave velocity, V_w is the acoustic velocity in water, and f is the acoustic frequency. The elastic modulus of the near-surface region of the specimen can be characterized from V_s:

2.4 Electronic speckle pattern interferometry (ESPI)

Figure 4a illustrates a typical electronic speckle pattern interferometry (ESPI) setup. A laser beam is split into two beams that constitute interferometric paths. The two beams are expanded and recombined on the surface of the specimen attached to the tensile machine. As superposition of coherent light beams, speckles

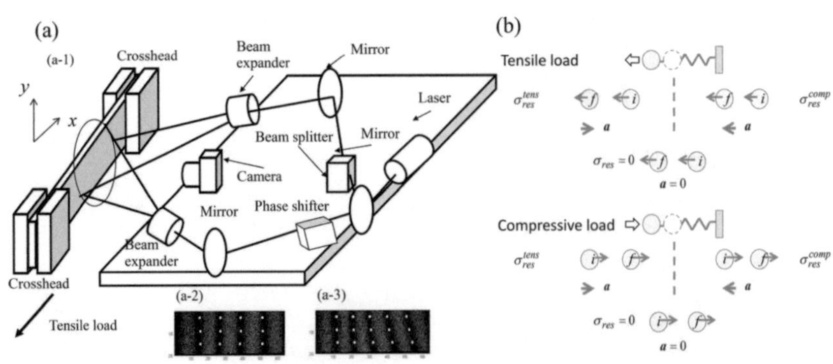

Figure 4.
(a) Typical ESPI setup. (a-1) Optical configuration. (a-2) Sample dark fringes. (a-3) Another sample dark fringes. (b) Principle of operation.

are formed in the image plane of a digital camera that captures images of the specimen. When the specimen is deformed by the tensile machine, the optical path length of one interferometric path increases and the other decreases. This changes the phase of each of all the speckles formed by the respective interfering beams. The digital camera takes images as the tensile machine keeps applying the load. The image captured at a certain time step is subtracted from the one captured at a different time step. The subtracted image exhibits an interferometric fringe pattern because the speckles undergo the phase change corresponding to the time difference between the two time steps. In those regions on the specimen where the speckles undergo a phase change corresponding to an integer multiple of 2π, the intensity of the interferometric images taken at the first and second time steps is the same. Consequently, dark fringes are formed, as illustrated by the inserts in **Figure 4a-2, a-3**.

In this arrangement the tensile machine applies a tensile load to the specimen so that the interference fringe patterns can be formed at least in three time steps. The applied load is kept as small as possible so that it does not relax the residual stress on the specimen. Since each of the fringe patterns contains the contours of displacement that the specimen undergoes in the duration between the first and second images are captured, the physical information contained by the fringe pattern represents the velocity. So, by subtracting two fringe patterns obtained by subtraction consecutively, the resultant frame contains acceleration of the points on the specimen surface. With the algorithm described below, it is possible to diagnose the status of residual stress through analysis of these frames containing the acceleration information.

Figure 4b illustrates the principle of operation. Consider that a certain part of the specimen has a compressive residual stress. As the top right part of this figure illustrates, the situation can be modeled by a mass connected to a compressed spring. If an external agent applies a tensile load (the load opposite to the compression), the mass returns to the equilibrium (represented by a dashed line in the figure) with acceleration in the same direction as the applied load. If the residual stress is tensile, the same external load displaces the mass away from the equilibrium. Hence, in this case, the acceleration of the mass is opposite to the applied load. This mechanism can be summarized as follows. "If the residual stress and applied load are of the same type (tensile or compression), the acceleration is opposite to the applied load. Otherwise, the acceleration is in the same direction as the applied load." The situation is the same when the applied load is compressive, as illustrated in the lower part of **Figure 4b**.

By applying this algorithm to diagnose the type of residual stress at a given point and combining it with the abovementioned formation of fringe pattern representing acceleration, it is possible to map out the type of residual stress on all points of the specimen.

It is also possible to estimate the elastic modulus at all points. As mentioned above, the fringe pattern resulting from subtraction of the interferometric image taken at one time step from another time step represents the displacement occurring in the time difference between the two time steps. By evaluating the displacement at all coordinate points (e.g., via interpolation between dark fringes) and dividing them by the pixel interval, it is possible to map out strain as a full-field two-dimensional data. By assuming that all the points on the specimen are under equilibrium when the tensile machine applies the load, the stress can be evaluated by dividing the applied load by the cross-sectional area of the specimen. This procedure yields a map of relative elastic modulus over the specimen. If the third-order elastic constant of the material is known, the elastic modulus can be calibrated by performing an acoustic velocity measurement at several points of the specimen.

This yields a map of absolute elastic modulus, and from this data the residual strain can be evaluated with Eq. (6).

Once the residual strain is found, the residual stress can be estimated as follows:

$$\sigma_{res} = \left(E_0 + C^{(3)}\varepsilon_{res}\right)\varepsilon_{res} \tag{11}$$

2.5 Finite element modeling (FEM)

Accurate numerical modeling of residual stress is extremely difficult. However, finite element modeling is useful because it can provide us with some insight and qualitative analysis when combined with experiment. This section discusses frameworks of such modeling in conjunction with the TOEC algorithm. For simplicity, the discussion here is limited to an isotropic case [24].

With the third-order term included, the constitutive relation can be expressed as follows:

$$\begin{pmatrix}\sigma_1\\ \sigma_2\\ \sigma_3\\ \sigma_4\\ \sigma_5\\ \sigma_6\end{pmatrix} = \left\{\begin{pmatrix} C_{11} & C_{12} & C_{13} & C_{14} & C_{15} & C_{16}\\ C_{21} & C_{22} & C_{23} & C_{24} & C_{25} & C_{26}\\ C_{31} & C_{32} & C_{33} & C_{34} & C_{35} & C_{36}\\ C_{41} & C_{42} & C_{43} & C_{44} & C_{45} & C_{46}\\ C_{51} & C_{52} & C_{53} & C_{54} & C_{55} & C_{56}\\ C_{61} & C_{62} & C_{63} & C_{64} & C_{65} & C_{66}\end{pmatrix} + \begin{pmatrix} \Delta C_{11} & \Delta C_{12} & \Delta C_{13} & \Delta C_{14} & \Delta C_{15} & \Delta C_{16}\\ \Delta C_{21} & \Delta C_{22} & \Delta C_{23} & \Delta C_{24} & \Delta C_{25} & \Delta C_{26}\\ \Delta C_{31} & \Delta C_{32} & \Delta C_{33} & \Delta C_{34} & \Delta C_{35} & \Delta C_{36}\\ \Delta C_{41} & \Delta C_{42} & \Delta C_{43} & \Delta C_{44} & \Delta C_{45} & \Delta C_{46}\\ \Delta C_{51} & \Delta C_{52} & \Delta C_{53} & \Delta C_{54} & \Delta C_{55} & \Delta C_{56}\\ \Delta C_{61} & \Delta C_{62} & \Delta C_{63} & \Delta C_{64} & \Delta C_{65} & \Delta C_{16}\end{pmatrix}\right\}\begin{pmatrix}\varepsilon_1\\ \varepsilon_2\\ \varepsilon_3\\ \varepsilon_4\\ \varepsilon_5\\ \varepsilon_6\end{pmatrix} \tag{12}$$

Here the first term inside the bracket on the right-hand side is the second-order elastic coefficient, and the second term is the third-order elastic coefficient. In the case of an isotropic material, the third-order elastic coefficient can be expressed as follows [25]:

$$\begin{pmatrix}\Delta C_{11}\\ \Delta C_{21}\\ \Delta C_{31}\\ \Delta C_{41}\\ \Delta C_{51}\\ \Delta C_{61}\end{pmatrix} = \begin{pmatrix} C_{111} & C_{112} & C_{112} & 0 & 0 & 0\\ C_{112} & C_{112} & C_{123} & 0 & 0 & 0\\ C_{112} & C_{123} & C_{112} & 0 & 0 & 0\\ 0 & 0 & 0 & C_{144} & 0 & 0\\ 0 & 0 & 0 & 0 & C_{155} & 0\\ 0 & 0 & 0 & 0 & 0 & C_{155}\end{pmatrix}\begin{pmatrix}\varepsilon_1\\ \varepsilon_2\\ \varepsilon_3\\ \varepsilon_4\\ \varepsilon_5\\ \varepsilon_6\end{pmatrix} \tag{13}$$

$$\begin{pmatrix} \Delta C_{12} \\ \Delta C_{22} \\ \Delta C_{32} \\ \Delta C_{42} \\ \Delta C_{52} \\ \Delta C_{62} \end{pmatrix} = \begin{pmatrix} C_{112} & C_{112} & C_{123} & 0 & 0 & 0 \\ C_{112} & C_{111} & C_{112} & 0 & 0 & 0 \\ C_{123} & C_{112} & C_{112} & 0 & 0 & 0 \\ 0 & 0 & 0 & C_{155} & 0 & 0 \\ 0 & 0 & 0 & 0 & C_{144} & 0 \\ 0 & 0 & 0 & 0 & 0 & C_{155} \end{pmatrix} \begin{pmatrix} \varepsilon_1 \\ \varepsilon_2 \\ \varepsilon_3 \\ \varepsilon_4 \\ \varepsilon_5 \\ \varepsilon_6 \end{pmatrix} \tag{14}$$

$$\begin{pmatrix} \Delta C_{13} \\ \Delta C_{23} \\ \Delta C_{33} \\ \Delta C_{43} \\ \Delta C_{53} \\ \Delta C_{63} \end{pmatrix} = \begin{pmatrix} C_{112} & C_{123} & C_{112} & 0 & 0 & 0 \\ C_{123} & C_{112} & C_{112} & 0 & 0 & 0 \\ C_{112} & C_{112} & C_{111} & 0 & 0 & 0 \\ 0 & 0 & 0 & C_{155} & 0 & 0 \\ 0 & 0 & 0 & 0 & C_{144} & 0 \\ 0 & 0 & 0 & 0 & 0 & C_{155} \end{pmatrix} \begin{pmatrix} \varepsilon_1 \\ \varepsilon_2 \\ \varepsilon_3 \\ \varepsilon_4 \\ \varepsilon_5 \\ \varepsilon_6 \end{pmatrix} \tag{15}$$

$$\begin{pmatrix} \Delta C_{14} \\ \Delta C_{24} \\ \Delta C_{34} \\ \Delta C_{44} \\ \Delta C_{54} \\ \Delta C_{64} \end{pmatrix} = \begin{pmatrix} 0 & 0 & 0 & C_{144} & 0 & 0 \\ 0 & 0 & 0 & C_{155} & 0 & 0 \\ 0 & 0 & 0 & C_{155} & 0 & 0 \\ C_{144} & C_{155} & C_{155} & 0 & 0 & 0 \\ 0 & 0 & 0 & 0 & 0 & C_{456} \\ 0 & 0 & 0 & 0 & C_{456} & 0 \end{pmatrix} \begin{pmatrix} \varepsilon_1 \\ \varepsilon_2 \\ \varepsilon_3 \\ \varepsilon_4 \\ \varepsilon_5 \\ \varepsilon_6 \end{pmatrix} \tag{16}$$

$$\begin{pmatrix} \Delta C_{15} \\ \Delta C_{25} \\ \Delta C_{35} \\ \Delta C_{45} \\ \Delta C_{55} \\ \Delta C_{65} \end{pmatrix} = \begin{pmatrix} 0 & 0 & 0 & 0 & C_{155} & 0 \\ 0 & 0 & 0 & 0 & C_{144} & 0 \\ 0 & 0 & 0 & 0 & C_{155} & 0 \\ 0 & 0 & 0 & 0 & 0 & C_{456} \\ C_{155} & C_{144} & C_{155} & 0 & 0 & 0 \\ 0 & 0 & 0 & C_{456} & 0 & 0 \end{pmatrix} \begin{pmatrix} \varepsilon_1 \\ \varepsilon_2 \\ \varepsilon_3 \\ \varepsilon_4 \\ \varepsilon_5 \\ \varepsilon_6 \end{pmatrix} \tag{17}$$

$$\begin{pmatrix} \Delta C_{16} \\ \Delta C_{26} \\ \Delta C_{36} \\ \Delta C_{46} \\ \Delta C_{56} \\ \Delta C_{66} \end{pmatrix} = \begin{pmatrix} 0 & 0 & 0 & 0 & 0 & C_{144} \\ 0 & 0 & 0 & 0 & 0 & C_{155} \\ 0 & 0 & 0 & 0 & 0 & C_{155} \\ 0 & 0 & 0 & 0 & C_{456} & 0 \\ 0 & 0 & 0 & C_{456} & 0 & 0 \\ C_{155} & C_{155} & C_{144} & 0 & 0 & 0 \end{pmatrix} \begin{pmatrix} \varepsilon_1 \\ \varepsilon_2 \\ \varepsilon_3 \\ \varepsilon_4 \\ \varepsilon_5 \\ \varepsilon_6 \end{pmatrix} \tag{18}$$

The isotropic TOE tensor is described by three linearly independent elements [24]. Choosing C_{123}, C_{144}, and C_{456} to be the three independent elements, we can use the following relations for complete expression of the third-order coefficient:

$$C_{111} = C_{123} + 6C_{144} + 8C_{456} \tag{19}$$

$$C_{112} = C_{123} + 2C_{144} \tag{20}$$

$$C_{155} = C_{144} + 2C_{456} \tag{21}$$

By substituting Eqs. (19)–(21) into Eqs. (13)–(18), we can find the third-order effect for each stress tensor component.

To compare with acoustoelastic measurement, it is necessary to express the effect of the inclusion of the third-order elastic coefficient in the corresponding acoustic velocity. Assuming that the density is unaffected by the inclusion of the third-order effect, the relative acoustic velocity can be expressed as follows:

$$\frac{v_{ij}^{(3)}}{v_{ij}^{(2)}} = \frac{\sqrt{\left(c_{ij} + \Delta C_{ij}\right)/\rho}}{\sqrt{C_{ij}/\rho}} = \sqrt{\frac{\left(c_{ij+\Delta C_{ij}}\right)}{C_{ij}}} \tag{22}$$

Here $i,j = 1\ldots6$, and $v_{ij}^{(3)}$ and $v_{ij}^{(2)}$ denote the acoustic velocity of the corresponding mode with and without the third-order effect.

3. Experimental observations and numerical analysis

3.1 Carbon steel and cemented carbide dissimilar welding

As an example of dissimilar welding, we discuss here a previous analysis [26] on butt-brazing of a carbon steel (skh51) and cemented carbide (V30). **Figure 5a** illustrates the arrangement of the brazing. An skh51 plate of 18.5 mm wide, 50 mm long, and 3.37 mm thick was placed on a mount with the 18.5 mm side contacting a V30 plate of the same dimension. For approximately 30 mm in length around the contacting area, an induction coil was arranged to heat the specimen for brazing. On brazing, Ag braze paste (Ag-Cu-Zn-Ni alloy, ISO Ag450) was put on the contact surface, and a braze temperature of 800°C was applied for 10 s. After this 10 s period, the brazed specimen was air-cooled. The noncontacting 18.5 mm sides of the respective plates were clamped to the steel blocks as shown in **Figure 5a**. The two steel blocks were connected via slide guides for free vertical slide. This means that the only constrain on the plates during the brazing operation was gravity. **Table 1** shows the material constants of skh51 and V30.

A contact acoustic transducer and a scanning acoustic microscope (SAM) were used for the analysis. The acoustic signal from the contact acoustic transducer travels through the entire thickness of the specimen. Thus, the measured acoustic velocity indicates the elastic property averaged over the specimen thickness. On the

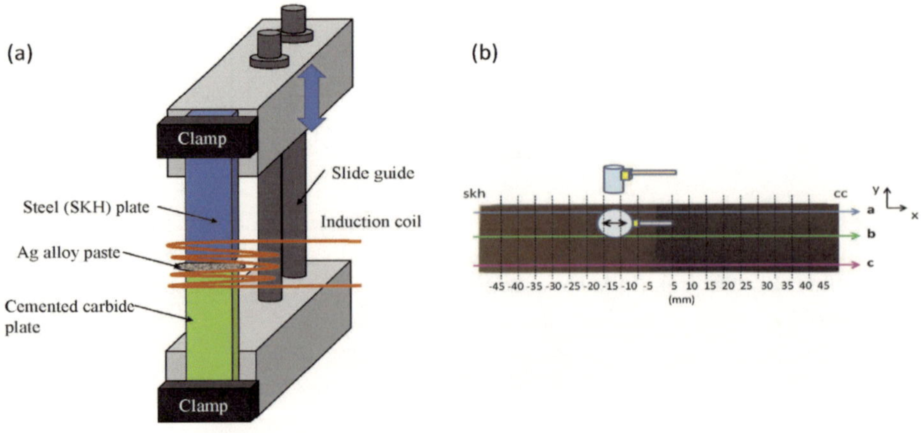

Figure 5.
(a) Setup for skh-cc welding (brazing). (b) Butt-brazed specimen and coordinate points for acoustic measurements.

Material	Steel (skh51)	Cemented carbide (V30)
Elastic modulus (GPa)	219	580
Thermal expansion (10^{-6} K^{-1})	11.9	5.3
Thermal conductivity (W m^{-1} K^{-1})	23.0	67.0

Table 1.
Material constants of the brazed plates.

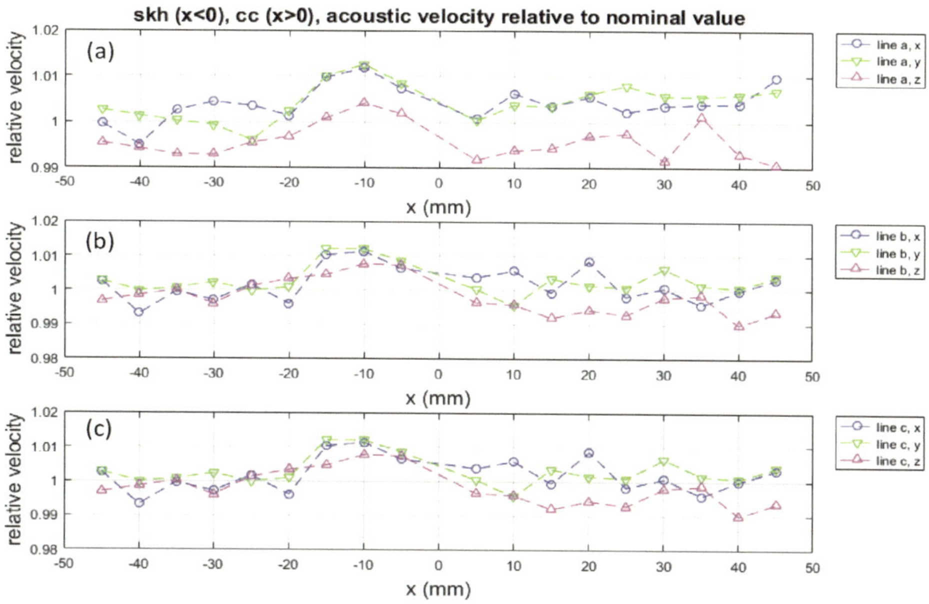

Figure 6.
Acoustic velocities relative to nominal value. (a) Line a, (b) line b, and (c) line c.

other hand, the acoustic signal emitted from the transducer head of the SAM is focused in a subsurface area of the specimen. Therefore, it detects the elastic property of the subsurface region. Hence, through a comparison of data from the contact acoustic transducer and the SAM, we can obtain information regarding the depth of the residual stress.

Figure 5b shows the butt-brazed specimen and the coordinate points where the acoustic measurements were conducted. The shear wave and longitudinal wave velocities were measured with contact acoustic transducers (Olympus V156-RM and M110-RM, respectively), driven commonly by a square wave pulser/receiver (Model 5077PR). The surface acoustic wave velocity was measured with a SAM (Olympus UH3) with 200 and 400 MHz transducer heads in the burst mode for V(z) curve analysis [22, 23].

3.2 Transverse residual stress profile

Figure 6 plots the acoustic velocities relative to the nominal values (measured before the brazing). Here the three graphs are for reference line a, line b, and line c (labeled in **Figure 5b**) from top to bottom. The following observations can be made.

3.2.1 Observation 1

The longitudinal wave (z-wave) velocity shows the following features. On the V30 side, it is lower than the nominal value for the entire horizontal span

($0 < x < 50$ mm). On the skh51 side, the velocity is clearly higher than the nominal value in the near-joint region ($-20 < x < 0$ mm) and slightly lower than the nominal value toward the cold end ($x < -20$ mm). These features can be translated into the following characteristics in the residual stress in the z-direction. The V30 side experiences tensile stress uniformly over the entire horizontal span. The skh51 side experiences compressive residual stress near the joint and very slight tensile stress for the rest of the horizontal span. Along reference line a, the situation toward the cold end on the skh51 side is slightly different. The material experiences clearer tensile residual stress.

3.2.2 Observation 2

The shear wave velocities (commonly for the x- and y-waves) show the following features. On the V30 side, the velocity is slightly higher than the nominal value for the entire horizontal span. On the skh51 side, the velocities are clearly higher than the nominal values in the near-joint region ($-20 < x < 0$ mm) and approximately the same as the nominal value for the rest of the region. These features indicate the following characteristics in the in-plane residual stresses. The V30 side experiences slight compressive residual stress uniformly over the horizontal span. On the skh51 side, the near-joint region experiences compressive residual stress. In the region toward the cold end, the material experiences tensile residual stress at a low level around reference lines b and c. Around reference line a, the residual stress in the x-direction is considerably compressive.

The above observations indicate the following overall residual stresses. On the V30 side, the residual stress is tensile along the thickness and slightly compressive along the surface plane. In both cases, the residual stress is uniformly distributed over the entire horizontal span. On the skh51 side, the residual stress is concentrated in the near-joint region where the residual stress is compressive in all directions. Toward the cold end, the residual stress is slightly tensile in all directions around reference lines b and c. Near reference line a, the residual stress is more tensile in the z-direction and compressive in the x-direction.

Possible explanations of these features found in the residual stresses are as follows:

1. The more uniform feature observed on the V30 side is due to the higher thermal conductivity (**Table 1**). The heat input from the joint flows relatively easily to the cold end on the V30 side. Consequently, the thermal effect is uniform on this size. Contrastively, due to the poor thermal conductivity, the heat input is confined in the near-joint region $-20 < x < 0$ mm (called the heat-affected zone, HAZ) on the skh51 side.

2. The compressive residual stress in the HAZ of skh51 ($-20 < x < 0$ mm) results from the following effects. In the heating phase, the HAZ experiences body-centered cubic (bcc) to face-centered cubic (fcc) phase transformation. This process is accompanied by an increase in the density. Consequently, the thermal expansion of the HAZ is smaller as compared with the non-heat-affected zone (the non-HAZ) in the $-50 < x < -20$ mm region. In the cooling phase, the HAZ of skh51 experiences fcc to bcc phase transformation. Consequently, it undergoes relatively smaller shrinkage.

3. The above phenomena affect the HAZ of the skh51side differently along the x-axis than the y- and z-axes. Along the x-axis, the following events take place. In the heating phase, the HAZ is compressed by the non-HAZ of the skh51 side and the V30 side. The non-HAZ undergoes larger thermal expansion because

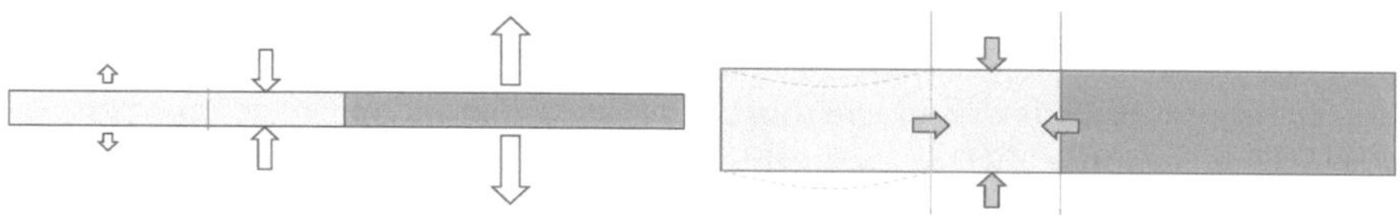

Figure 7.
Deformation induced by thermal load due to brazing and other constraints. Inward arrows represent compressive residual stress and outward arrows tensile residual stress. Sizes of arrows represent the magnitude of residual stress.

the phase transformation does not take place. The V30 side experiences uniform thermal expansion with the cool end constrained by the table of the welding setup. The gravity acts in favor of this compression experienced by the HAZ. The greater elastic modulus of V30 also helps this compressing mechanism. In the cooling phase, the HAZ of skh51 shrinks less than the other regions (the non-HAZ of skh51 and the V30 side). This makes the HAZ of skh51 tend to be stretched by the other regions. However, this time the stretching force is against the gravity. Consequently, the compressive stress formed in the heating phase remains in the HAZ on the skh51 side.

4. The events along the y- and z-axes are slightly different. At the boundary (the joint), the higher thermal conductivity makes the expansion (in the heating phase) and shrinkage (in the cooling phase) faster on the V30 side than the skh51 side. Consequently, when the skh51 side is still shrinking, the V30 side has already completed the shrinkage, preventing the skh51 side from further shrinkage. Consequently, the skh51 side is compressed along the y- and z-axes at the boundary.

5. The residual stress on the V30 side results from the above events. As the reaction to the compressive stress on the skh51 in the cooling phase, the V30 side experiences tensile stress in the y and z-directions at the boundary. Between these two directions, the material is less constrained in the z-direction due to the shorter span of the joint. Along the x-axis, the V30 side is constrained by the skh51 at the joint and by the table at the cool end. Consequently, it is likely that the tensile deformation occurs preferably in the z-direction. By Poisson's effect, the material on the V30 side undergoes compressive deformation in the x and y-directions. This explains the observation of relative acoustic velocity greater than unity in these directions on the V30 side.

6. The different behavior observed on the skh51 side along reference line a from reference lines b and c is due to the involvement of rotational displacement at the boundary. When the V30 side exerts compressive force in the x-direction on the skh51 side at the joint, the force induces counterclockwise rotation around the z-axis (in **Figure 5b**). This causes the compressive stress on the reference line a side of the specimen.

Figure 7 illustrates this observation schematically with some exaggeration.

4. Depth of residual stress

Another important factor in residual stress analysis in general is the information regarding the depth of residual stresses. The capability to detect the depth of residual stresses is important not only to identify the depth of a given residual stress accurately but also to prevent the possibility to overlook residual stresses. The X-ray diffractometry (XRD) is one of the most developed methods of residual stress analysis. However, the XRD is applicable to the identification of residual stresses within the penetration depth of the X-ray, which is typically a few hundreds of μm from the surface.

One way to evaluate the depth of residual stresses is to apply a contact acoustic transducer and SAM to the same specimen. In the experiment discussed here [27], the following configurations are used for the acoustic devices: (i) the SAM (Olympus UH3) with a 400 MHz transducer head and a spherical lens (310 μm focal length); (ii) the SAM with a 200 MHz transducer head and the same spherical lens as (ii); (iii) the SAM with a 200 MHz transducer head and a cylindrical lens of the same focal length (310 μm) as the spherical lens; (iv) the contact acoustic transducer with a longitudinal wave sensor head (M110-RM); and (iv) the contact acoustic transducer with a shear wave sensor head (V156-RM). These configurations work differently as follows. The SAM generates a Rayleigh wave [28] on the surface of the specimen [23]. The Rayleigh wave is mostly out of plane (normal to the specimen surface) but contains in-plane components (the component whose acoustic oscillation is parallel to the surface). When the incident acoustic wave is focused with a spherical lens, a surface acoustic wave is generated in random directions parallel to the specimen surface. Consequently, the in-plane components are averaged out, and the resultant (superposed) acoustic oscillation is out of plane. So, signals with configurations (i), (ii), and (iv) are sensitive to the elastic modulus in the z-direction. When the same incident wave from the SAM is line-focused with a cylindrical lens, on the other hand, the in-plane components whose acoustic oscillation is perpendicular to the axis of the cylindrical lens are not averaged out. Thus, the resultant acoustic wave is sensitive to the elastic modulus in the direction perpendicular to the lens's axis. So, signals with configurations (ii) and (v) are sensitive to the elastic modulus in the x or y direction (depending on the orientation of the cylindrical lens and the shear wave sensor head).

Figure 8 compares the measurement conducted with configurations (i) SAM 400 MHz spherical lens, (ii) SAM 200 MHz spherical lens, and (iv) contact acoustic transducer with the longitudinal wave mode. The 400 MHz SAM and 200 MHz SAM used the same acoustic spherical lens (310 μm focal length). Hence, the angle of incidence of the acoustic wave to the specimen surface was the same. The penetration depth of the acoustic wave is of the order of the acoustic wave length, i.e., approximately 15 μm for skh51 and 17 μm for V30. The longitudinal bulk acoustic wave from the contact transducer passes through the specimen and reflects off the rear surface. Therefore, the acoustic velocity data represents the elastic

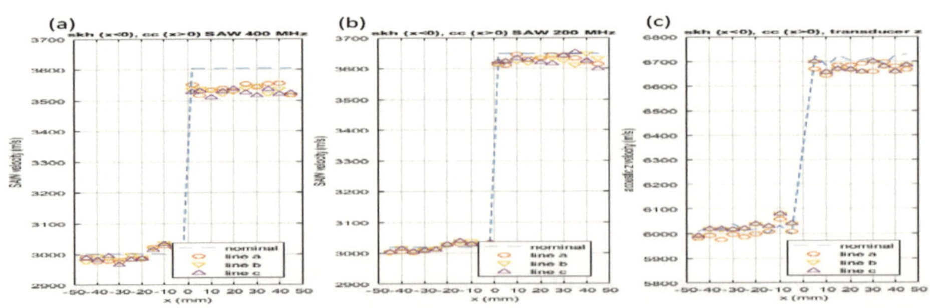

Figure 8.
Comparison of acoustic velocity data obtained with (a) SAM with 400 MHz head, (b) SAM with 200 MHz head, and (c) contact acoustic transducer.

modulus averaged over the entire thickness. The acoustic velocity data with 400 MHz SAM shows much greater deviation in the acoustic velocity from the nominal value than the 200 MHz SAM or the contact transducer. On the other hand, the data with (ii) the 200 MHz SAM and (iv) the contact transducer with the longitudinal mode are similar to each other. **Figure 9** shows the similarity between (ii) and (iv) more clearly by expressing the signal and the relative velocity to the nominal value. These indicate that the residual stress in this specimen is localized within approximately 20 μm from the surface.

While the longitudinal wave data taken with the contact acoustic transducer is similar to the SAM data with 200 MHz, the shear wave data obtained with (v) the contact acoustic transducer shows considerable difference from the data taken with (iii) the 200 MHz SAM with the cylindrical lens of the same focal length (310 μm focal length) as the spherical lens. **Figure 10** compares the data between the contact acoustic transducer and SAM for the oscillation direction of x and y, respectively. In both directions, the signal from the SAM shows considerable difference from the bulk shear wave from the contact acoustic transducer on the V30 side, while no difference is seen on the skh51 side. This indicates that on the V30 side, the material undergoes significant in-plane tensile deformation at the region twice as deep as the out-of-plane expansion (the wavelength at 200 MHz is twice of that at 400 MHz).

Comparison was also made between the optical interferometric data and acoustic data. **Figure 11** compares the acceleration map obtained by the ESPI and relative acoustic velocity (shear wave with oscillation along x). Qualitative agreement is seen indicating the consistency between the stress evaluation based on the acceleration algorithm and the acoustoelasticity.

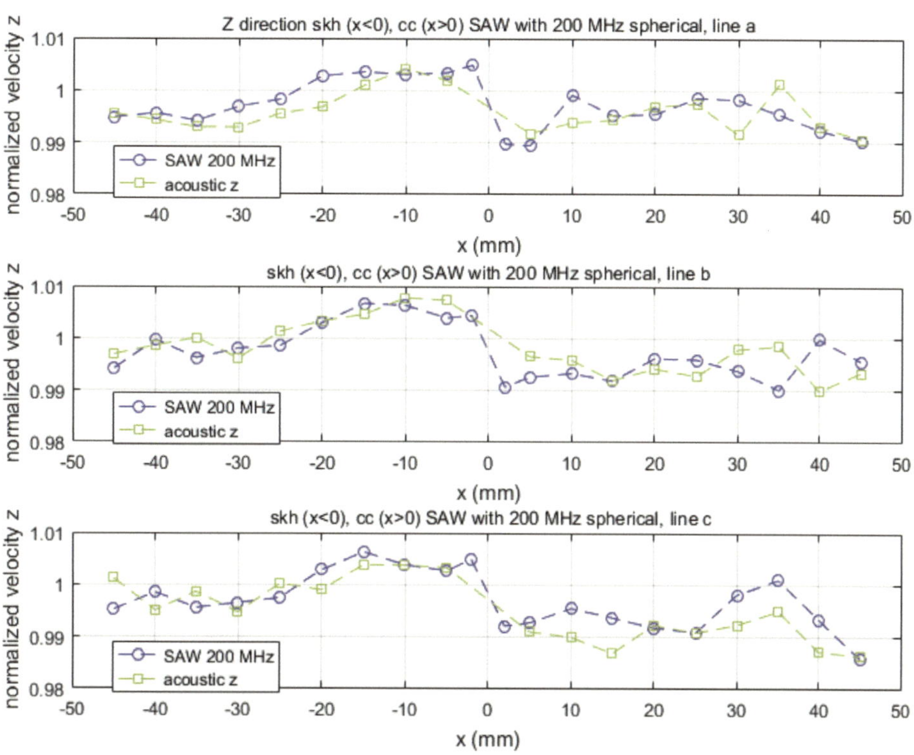

Figure 9.
Comparison of SAM with 200 MHz head data and contact acoustic transducer.

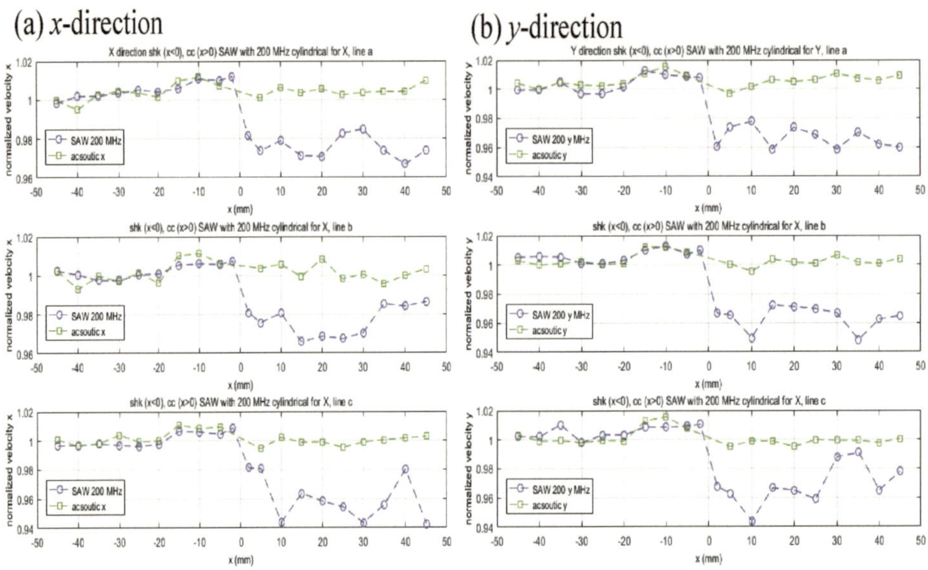

Figure 10.
Comparison of shear wave velocity obtained with contact acoustic transducer and SAM with 200 MHz shear mode transducer head.

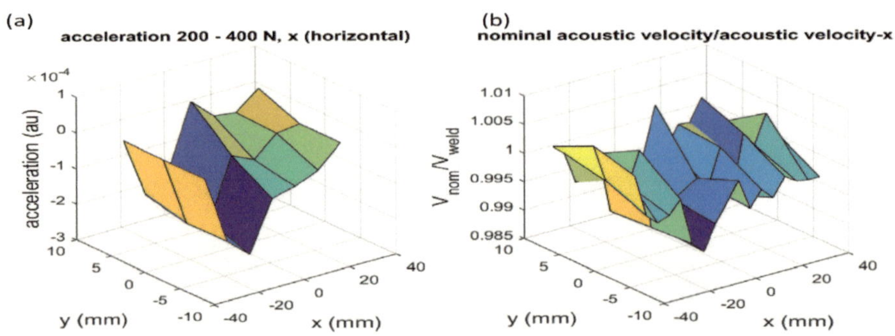

Figure 11.
Comparison of ESPI acceleration (x) and acoustic velocity (x). (a) ESPI, and (b) Acoustic transducer x.

4.1 Bead-on-plate welding on Al 5083

A similar analysis was made for an aluminum alloy specimen [29]. The specimen used in this experiment was a bead-on-plate aluminum alloy 5083 [30] processed by gas tungsten arc (TIG) welding. **Figure 12** shows a photo of the specimen and the steel block used to place the specimen for welding. **Table 2** lists the welding condition. The specimen plate was placed on a steel block that held one end of the specimen without imposing any other constraint. The TIG welding torch applied

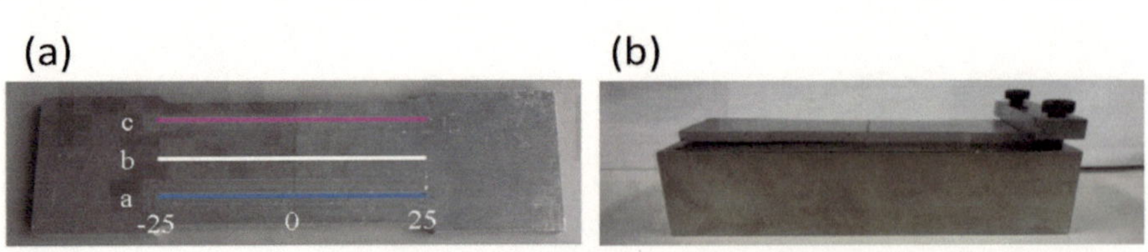

Figure 12.
(a) Al 5083 bead-on-plate specimen. Acoustic measurements were made along line a-c. (b) Specimen placed on steel block for welding.

Welding speed	Welding current	Welding voltage	Shielding gas	Cooling
5.0 mm/s	100 A	200 V	Ar (10 l/min)	Ambient air

Table 2.
Bead-on-plate welding condition.

the heat on the top surface of the specimen from the bottom end (the side close to reference line a in **Figure 12**) toward the top end (the side close to line c). During the heating process and the postheating phase, the specimen was air-cooled.

5. Acoustic, optical, and XRD measurements

For the Al 5083 specimen, the same type of acoustic and optical measurement as the skh51-V30 dissimilar weld specimen was made. The same contact acoustic transducer was used at the same coordinate points for the acoustic wave velocity measurement, and the same ESPI setup was used for the acceleration measurement. In addition, XRD analysis based on the 2θ- $\sin^2\Psi$ plot and FEM analysis were conducted. The FEM model simulated the TIG welding by a Gaussian profiled heat input (1.2 cm full width at half maximum) moving at the speed of the welding torch with an electric power of 20 kW and a coupling coefficient to the specimen of 0.9%. The cooling phase (500 s) was simulated with natural convection at all the surfaces. The deformation was made permanent when the strain exceeded a preset yield strain. More about this modeling can be found in [29]. **Figure 13** compares the results from the above four types of analyses. The data is presented in the form of two-dimensional mapping where the vertical axis represents a physical quantity associated with the residual stress in the x-direction. The ESPI analysis presents the acceleration in the x-direction, the XRD and FEM analyses present x-component of

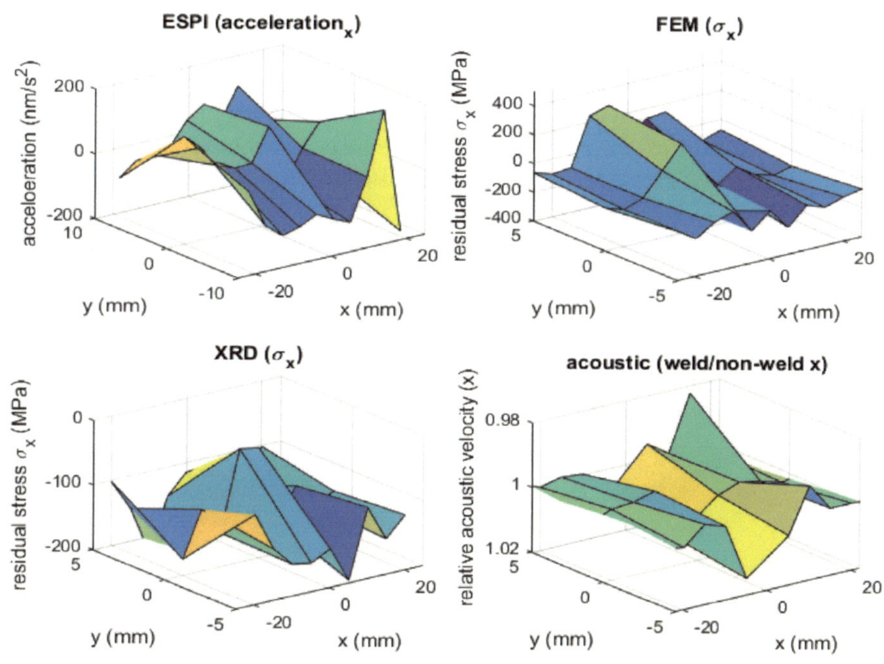

Figure 13.
Comparison of ESPI, XRD, FEM, and acoustic results.

the residual stress, and the acoustic result presents the relative acoustic velocity (normalized to the nominal velocity) of the x-shear wave.

Although details are different, the results from all the four analyses show qualitative agreement in the following sense. (1) All data show the general tendency that the central (near weld) region tends to have tensile residual stresses and the boundary between the central and outside regions has more compressive residual stresses. (2) Along the weld line, all data indicate that the end point of welding has higher tensile stress than the start point of welding. These observations are consistent with the following widely accepted explanation. When the welding torch heats the work, the region near the weld (the central region) is thermally expanded pushing the sides of the weld line toward the ends of the specimen (sideways). The outside regions are cooler than the central region. Consequently, the outside regions experience less thermal expansion and constrain the thermal expansion of the central region. This causes a compressive residual stress at the boundary between the central and outside regions.

In the cooling phase, the higher heat dissipation on the bottom surface locks the abovementioned stress pattern on the top surface. As the temperature of the central region goes down, the material in this region tries to shrink. However, the region behind this central region (the region closer to the bottom surface) is relatively cooler from the beginning and, therefore, does not shrink as much as the top surface. This prevents the central region near the top surface from shrinking back to the initial length. In other words, it locks tensile residual stress near the top surface.

5.1 Effect of heat dissipation

According to the above arguments, the tensile residual stress of the central region is determined by the differential rate of heat dissipation between the top and bottom surface. It is possible that increases in heat dissipation from the bottom surface (the non-welded side) can reduce this tensile stress (because it takes the heat near the top surface more aggressively). A numerical study was made to test this scenario, and the result is presented in **Figure 14**. **Figure 14a** is the case when the same convection rate is used for the top and bottom surfaces. **Figure 14b** is the case when the convection on the bottom surface is increased by a factor of two. It is seen that when the heat dissipation from the bottom surface is higher, the tendency is that the central region along the weld line is reduced. This observation confirms that in the abovementioned scenario, the heat dissipation from the bottom surface

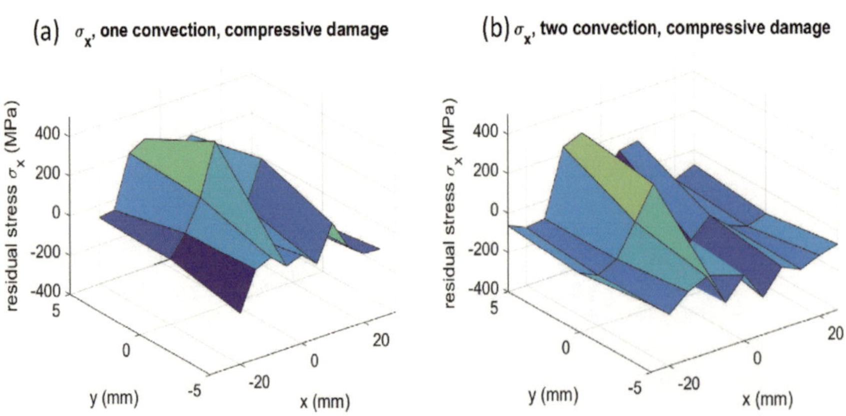

Figure 14.
Effect of heat dissipation rates. (a) The same convection is used for top and bottom surfaces; (b) convection is doubled for bottom surface.

causes the tensile residual stress on the top surface in the following sense. If the heat dissipation from the bottom surface is higher, the temperature gradient along the thickness of the specimen is lower in the cooling phase. This reduces the above-argued locking mechanism.

The case when the bottom surface has higher heat dissipation rate appears to be closer to the experimental result with ESPI shown in **Figure 13**. This is not surprising because the specimen was placed on a steel block (**Figure 12**) that behaved as a heat sink through the bottom surface. Contrastively, the top surface was exposed to air. It is reasonable to assume that the steel block was a greater heat sink than the ambient air.

5.2 Effect of permanent strain setting

The present FEM simulates the residual stress by making the strain exceed the yield strain permanent. As discussed above, during the heating and cooling phases, both tensile and compressive residual stresses are created. This indicates that both tensile and compressive permanent strain can create residual stress, and it is a good question which type has greater contribution to the creation of residual stress. A numerical study was conducted to investigate this effect.

Figure 15 shows the x-component of residual stress on the top surface. Here the x-axis is the axis parallel to the top surface and perpendicular to the weld line. **Figure 15a–c** shows the results from the three different conditions regarding the permanent deformation setting. (a) is the case when only the compressive yield strain is used for the permanent deformation setting, (b) is the case when both of the compressive and tensile yield strains are used, and (c) is the case when the tensile yield strain only is used. The heat dissipation setting for these cases is that the bottom surface has twice as high convection than the top surface. **Figure 15a, b** shows practically no difference. On the other hand, (c) is different from (a) and (b). These altogether indicate that the effect of compressive permanent deformation plays a more important role than the effect of tensile permanent deformation in the formation of residual stresses.

5.3 Effect of TOEC

Acoustoelastic method relies on the nonlinear elasticity (the third-order elastic constant). It is interesting to compare the acoustic velocity measured for all three

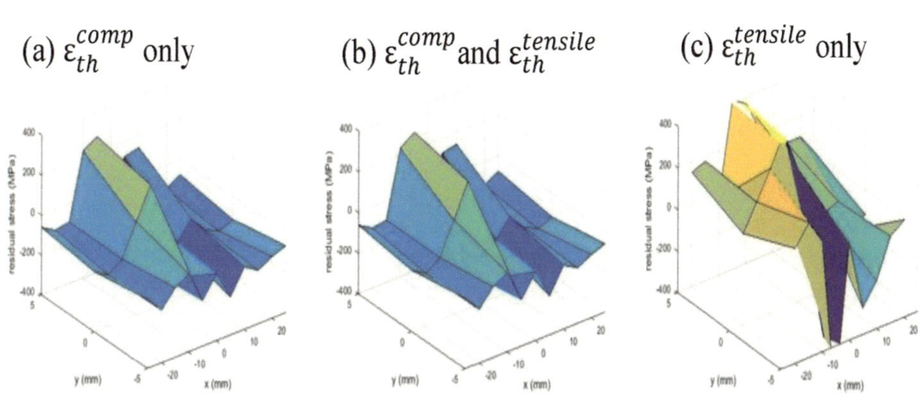

Figure 15.
Effect of permanent deformation setting. (a) Compressive yield strain only, (b) compressive and tensile yield strain, and (c) tensile yield strain only is considered.

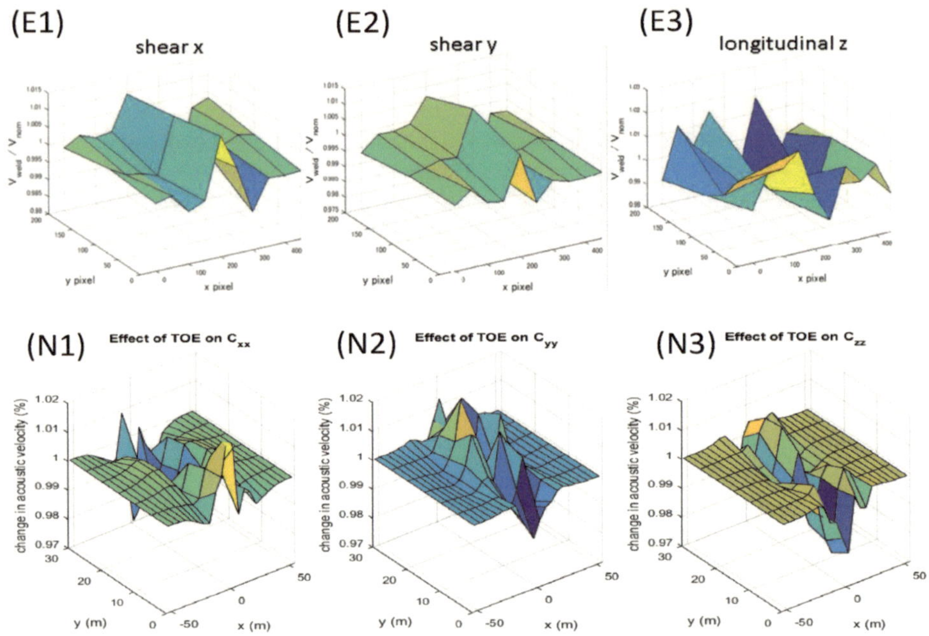

Figure 16.
Effect of TOEC based on the comparison of experimental relative acoustic velocity data. (E1)–(E3): Experimental relative velocity in x, y, *and* z*-directions. (N1)–(N3): Numerical data corresponding to (E1)–(E3).*

directions with numerical analysis conducted with the finite element model discussed above.

The top rows of **Figure 16** are the relative acoustic velocity (acoustic velocities normalized to the nominal value) measured with the contact acoustic transducer. The bottom graphs are numerical data corresponding to the three relative acoustic velocities, evaluated with the use of Eq. (22). For all three directions, the experimental and numerical results show qualitative agreement in the overall shape of the graphs. More importantly, **Figure 16** shows that the experimental and numerical changes in the acoustic velocity are in the same range, i.e., compressive/tensile residual stresses increase/decrease the acoustic velocity approximately by 1–2%. It should be noted that the root-mean-square error in the experimental acoustic velocity measured on a specimen with no residual stress is less than 0.05%.

The above agreement opens up a new way to use the conventional acoustoelastic technique. The general procedure is as follows. It is assumed that the second-order and third-order elastic coefficients (c_{ij}, c_{ijk}, etc.) are known:

Step 1: Measure the change in the acoustic velocity for all degrees of freedom, i.e., the longitudinal wave along the x (y, z) axis and the shear wave along the xy (yz, zx) surface.

Step 2: Use Eq. (22) backward to compute ΔC_{ij}.

Step 3: Use Eq. (13), etc. to find the strain vector.

Step 4: Use Eq. (11) to find the residual stress.

This procedure has not been tested yet but is a subject of our future study.

6. Summary

Applications of multiple methods to residual stress analyses are discussed. The idea behind the use of multiple methods is to compensate drawbacks of each method. Previous experimental and numerical studies on welding-induced residual stresses are presented and discussed. These previous results show consistency among the different methods used. Further research is definitely necessary to confirm more quantitative agreement in the results obtained by the different methods. Some procedural proposals are presented as subjects of future research. It is our hope that the content of this article is helpful to engineers and researchers of the related fields.

Acknowledgements

The present research is in part supported by the Louisiana Board of Regents Pilot-funded grant LEQSF(2016-2017)-RD-C-13 and Southeastern Louisiana University STAR grant. The authors are grateful to Robert P. Waldron Jr. for his assistance in this project.

Author details

Sanichiro Yoshida[1*] and Tomohiro Sasaki[2]

1 Department of Chemistry and Physics, Southeastern Louisiana University, Hammond, LA, USA

2 Department of Mechanical Engineering, Niigata University, Niigata, Japan

*Address all correspondence to: syoshida@selu.edu

References

[1] Schajer GS, editor. Practical Residual Stress Measurement Methods. Chichester, UK: Wiley; 2013

[2] Kandil FA, Lord JD, Fry AT, Grant PV. A Review of Residual Stress Measurement Methods—A Guide to Technique Selection; NFL Report MATC(A)O4. Teddington, UK: National Physical Laboratory; 2001

[3] Schajer GS, Yang L. Residual-stress measurement in orthotropic materials using the hole-drilling method. Experimental Mechanics. 1994;**12**:324-333

[4] Eijpe MPIM, Powell PC. Determination of residual shear stress in composite by a modified layer-removal method. Journal of Materials Science. 1998;**33**:20192026

[5] Roberts SJ, McCartney LN, Broughton WR. A novel method of determining stress distribution in plates using the incremental slitting technique. Journal of Strain Analysis for Engineering Design. 2011;**46**:280-297

[6] Tebedge N, Alpsten G, Tall L. Residual-stress measurement by the sectioning method. Experimental Mechanics. 1973;**13**:88-96

[7] Drescher-Krasicka E, Willis JR. Mapping stress with ultrasound. Nature. 1996;**384**:52-55

[8] Drescher-Krasicka E, Ostertag CP. Residual stress measurements in welded steel beam columns by scanning acoustic microscopy. Journal of Materials Science. 1999;**34**:4173-4179

[9] Fitzpatrickl ME, Fry AT, Holdway P, Kandil FA, Shackleton J, Suominen L. Determination of residual stress by X-ray diffraction—Issue 2. In: Measurement Good Practice Guide. Vol. 52. Teddington, UK: National Physical Laboratory; 2005

[10] Fitzpatrick ME, Lodini A. Analysis of Residual Stress by Diffraction Using Neutron and Synchrotron Radiation. New York, USA: CRC Press; 2003

[11] Hughes DS, Kelly JL. Second-order elastic deformation of solids. Physics Review. 1953;**92**:1145-1149

[12] Duquennoy M, Ouaftouh M, Ourak M, Jenot F, Xu WJ. Theoretical and experimental determination of the natural and initial acoustoelastic coefficients. Review of Progress in Quantitative Nondestructive Evaluation. 2002;**21**:16961702

[13] Dorfi HR, Busby HR, Janssen M. Ultrasonic stress measurements based on the generalized acoustic ratio technique. International Journal of Solids and Structures. 1996;**33**:11571174

[14] Duquennoy M, Ouaftouh M, Ourak M, Jenot F. Theoretical determination of Rayleigh wave acoustioelastic coefficients: Comparison with experimental values. Ultrasonics. 2002;**39**:575583

[15] Berruti T, Gola MM, Briggs GD. Acoustoelastic measurements on aluminum alloy by means of a contact and non-contact (LFB acoustic microscopy) technique. Journal of the Acoustical Society of America. 1998;**103**: 1370-1376

[16] Lee YC, Kuo SH. A new point contact surface acoustic wave transducer for measurement of acoustoelastic effect of polymethylmethacrylate. IEEE Transactions on Ultrasonics, Ferroelectrics, and Frequency Control. 2004;**51**:114-120

[17] Muir DD. One-sided ultrasonic determination of third order elastic constants using angle-beam acoustoelasticity measurements

[PhD thesis]. Atlanta, GA, USA: Georgia Institute of Technology; 2009

[18] Vikram CS, Pechersky MJ, Feng C, Engelhaupt D. Residual-stress analysis by local laser heating and speckle-correlation interferometry. Experimental Techniques. 2008;**20**:2730

[19] Withers PJ. Depth capabilities of neutron and synchrotron diffraction strain measurement instruments. II. Practical implications. Journal of Applied Crystallography. 2004;**37**: 607-612

[20] Casavola C, Campanelli SL, Pappalettere C. Preliminary investigation on distribution of residual stress generated by the selective laser melting process. Journal of Strain Analysis for Engineering Design. 2009;**44**:93104

[21] Prevéy PS. X-ray diffraction residual stress techniques. In: Metals Handbook. Vol. 10. Metals Park: American Society for Metals; 1986. pp. 380-392

[22] Zinin PV. Quantitative acoustic microscopy of solids. In: Levy M, Bass H, Stern R, Keppens V, editors. Handbook of Elastic Properties of Solids, Liquids, and Gases. Vol. I: Dynamic Methods for Measuring the Elastic Properties of Solids. New York, USA: Academic Press; 2001. pp. 187-226

[23] Zinin PV, Arnold W, Weise W, Berezina S. Chapter 11: Theory and applications of scanning acoustic microscopy and scanning near-field acoustic imaging. In: Kundu T, editor. Ultrasonic and Electromagnetic NDE for Structure and Material Characterization: Engineering and Biomedical Applications. New York, USA: CRC Press; 2012

[24] Fuck RF, Tsvankin I. Analysis of the symmetry of a stressed medium using nonlinear elasticity. Geophysics. 2009; **74**(5):WB79-WB87

[25] Barsch GR, Chang Z. Second-and higher-order effective elastic constants of cubic crystals under hydrostatic pressure. Journal of Applied Physics. 1968;**39**:3276-3284

[26] Yoshida S, Sasaki T, Usui M, Sakamoto S, Gurney D, Park IK. Residual stress analysis based on acoustic and optical methods. Materials. 2016;**9**(2):112. DOI: 10.3390/ma9020112

[27] Yoshida S, Sasaki T, Usui M, Sakamoto S, Park I-K, Jung JH, et al. Near weld stress analysis with optical and acoustic methods. In: Bossuyt S, Schajer G, Carpenteri A, editors. Residual Stress, Thermomechanics and Infrared Imaging, Hybrid Techniques and Inverse Problems, Vol. 9, Conference Proceedings of the Society for Experimental Mechanics; 2016. pp. 119-127

[28] Graff KF. Wave Motion in Elastic Solids. New York, USA: Dover; 1975

[29] Yoshida S, Miura F, Sasaki T, Didie DR, Rouhi S. Optical analysis of residual stress with minimum invasion. In: Baldi J, Considine M, Quinn S, Balandraud X, editors. Residual Stress, Thermomechanics and Infrared Imaging, Hybrid Techniques and Inverse Problems, Vol. 8, Conference Proceedings of the Society for Experimental Mechanics; 2018. pp. 65-72

[30] Aluminum Alloys Aluminum 5083 Properties, Fabrication and Applications. Supplier Data by Aalco. Available from: http://www.azom.com/article.aspx?ArticleID=2804 [Accessed: 13 March 2019]

Chapter 3

Nondestructive Characterization of Residual Stress Using Micromagnetic and Ultrasonic Techniques

Madalina Rabung, Meisam Amiri, Michael Becker, Melanie Kopp, Ralf Tschuncky, Ines Veile, Fabian Weber, Miriam Weikert-Müller and Klaus Szielasko

Abstract

Material processing and service loads in different lifecycle stages of a product—ranging from semi-finished goods to operating structures—lead to an unfavorable superposition of residual stresses, especially of micro- and macro-residual stresses. Whereas near-surface compressive stress is often desired as it prolongs the useful service life, undesired, steep stress gradients and tensile stress at the surface promote the occurrence of cracks and wear during operation, ultimately leading to expensive and possibly dangerous premature component failure. Reliable management of the residual stress condition significantly contributes to the assessment and optimization of a part's or component's lifetime. Therefore, the nondestructive evaluation of residual stress in objects of different scales reaching from laboratory samples over semi-finished products up to operating components and structures has gained significant importance in the latest decades. Micromagnetic and ultrasonic methods are based on the interaction of an external magnetic field or an ultrasonic wave, respectively, with the material's microstructure and residual stress fields on different scales and in different depths from the material surface. The present contribution provides an overview regarding the local and volumetric measurement, characterization and evaluation of macro- and micro-residual stress by means of micromagnetic and ultrasonic techniques.

Keywords: residual stress, nondestructive, micromagnetic, ultrasound, EMAT

1. Introduction

Residual stress is the totality of internal forces per area acting upon a boundary surface of a volume within a material, thereby maintaining the current size and shape of a given object under the absence of external loads [1, 2]. Depending on the range of constant stress value and direction, residual stress in polycrystals is divided into first kind (constant across larger distance, e.g., across several crystals/grains), second kind (constant across smaller distance, e.g., within a crystal/grain) and third kind (varies on the submicroscopic scale, e.g., from atom to atom) [2].

Although best expressed as a tensor, industrial practice often demands for simplified residual stress values regarding one or a few relevant directions, such as those most critical under the expected service load.

Residual stress has significant impact on product quality and useful service life, as it affects the product geometry and processability (strip steel being just one example), the tendency to crack appearance (e.g., after hardening or forming) as well as the fatigue lifetime. The present contribution discusses micromagnetic and ultrasonic techniques that are industrially applied for nondestructive stress measurement.

2. Theoretical background on the used nondestructive methods

2.1 Micromagnetic techniques

2.1.1 Magnetic structure

The prerequisite for the application of micromagnetic nondestructive testing methods is a ferromagnetic behavior of the material to be tested. Such ferromagnetic materials have a characteristic domain structure. Here, the domains represent areas spontaneously magnetized up to saturation, which are separated from each other by the so-called Bloch walls [3]. The alignment of the magnetization vectors of the domains is statistically distributed over a ferromagnetic sample in the non-magnetized state in such a way that their total effect is completely neutralized. The magnetic structure of iron-based ferromagnetic materials consists of two kinds of Bloch walls, 180° Bloch walls and 90° Bloch walls. The indicated angle represents the angle between the magnetization vectors in two neighboring domains. The 180° Bloch walls have short-range stress fields resulting in their strong interaction with microstructure inhomogeneity situated locally (dislocations or precipitates) or micro-residual stresses. The 90° Bloch walls have long-range stress fields that cause a strong interaction with microstructure heterogeneities stretched out over hundreds of microns and more (microstructure phases) or residual stresses of the second or first kind.

Under the influence of an external magnetic field, ferromagnetic materials align the magnetization vectors of the domains in the direction of the external magnetic field. The domains also change the structure under the influence of stresses. Known processes, Bloch wall movements and rotations of the domain magnetization vectors, cause these changes. The underlying effect is magnetostriction and the Villari effect. Magnetostriction describes the change in length of a ferromagnetic material during magnetization. A distinction is made between materials with positive and negative magnetostriction, depending on whether a sample lengthens or shortens in the field direction. The magnitude of the change in length depends on the direction of the magnetic field relative to the crystal axes. The Villari effect describes the reversal case that an elastic change in geometry causes a change in magnetization state. The cause of the effects lies in the spin-orbit coupling [4, 5].

2.1.2 Magnetic hysteresis

With an external alternating field magnetization of ferromagnetic materials, there is no clear correlation between the magnetic field strength H and the magnetic flux density B. This means that the magnetization of a ferromagnetic material depends not only on the external field strength (H) but also on the time course of the magnetization [4, 5]. If the magnetic field strength H is increased in a non-magnetized ferromagnetic material, the magnetic flux density B increases in its surroundings. If the field strength is reduced to zero again, the magnetic flux density remains at a value unequal to zero. The ferromagnetic material retains some residual magnetism (remanence). With periodic magnetization, a ferromagnetic material reacts with a hysteretic relation of flux density B and magnetic field H (see **Figure 1**) [4, 5]. With continuous periodic magnetization, the hysteresis curve is traversed once with each magnetization period.

The reason for the occurrence of hysteretic behavior is the interaction of Bloch walls and microstructure. The Bloch wall motion is discontinuous throughout the magnetization process, as the Bloch walls are pinned by existing lattice defects, and each pinning has to be overcome by increasing applied magnetic field. Lattice defects in the microstructure such as grain boundaries, dislocations and precipitates represent energetic minima for the Bloch walls. This causes the Bloch walls to tear loose abruptly and move until a pinning site of higher strength is reached.

Stresses affect the magnetic hysteresis behavior of ferromagnetic materials. In magnetostrictively positive materials (such as most ferromagnetic steels), tensile stress causes an increase of the differential susceptibility and a narrower hysteresis (H_C shift toward smaller values) and vice versa for compressive stresses. Magnetostrictively negative materials behave contrary to positive materials [4, 5, 7].

2.1.3 Magnetic Barkhausen noise

The analysis of magnetic Barkhausen noise (MBN) is based on a discovery by Barkhausen [8]: Abrupt changes in local magnetization caused by the discontinuous Bloch wall motion can be picked up inductively as a series of voltage pulses (noise) induced into a coil that is placed close to the material surface. The location of the highest density of these impulses is in relation to the magnetic hysteresis near the coercive force H_C, since this is where most Bloch wall jumps occur. The voltage pulses can be amplified, filtered and rectified for evaluation. The envelope of the filtered and rectified noise signal can be recorded as a function of the magnetization field strength, resulting in a so-called Barkhausen noise profile curve M(H) (see **Figure 2**). The magnetic Barkhausen noise has a characteristic dependence on

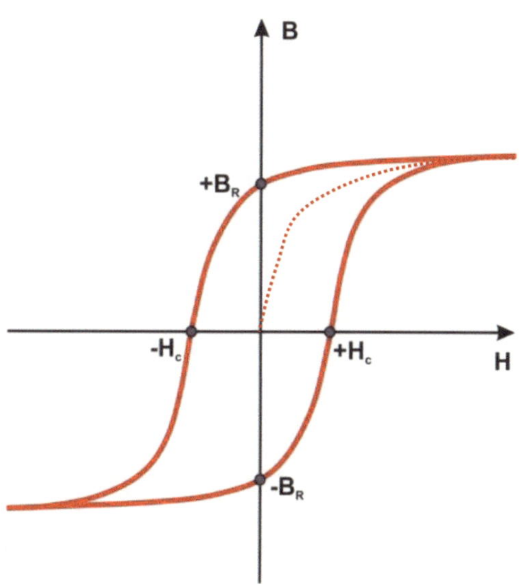

Figure 1.
Hysteresis curve (H_C, coercivity; B_R, remanence) [6].

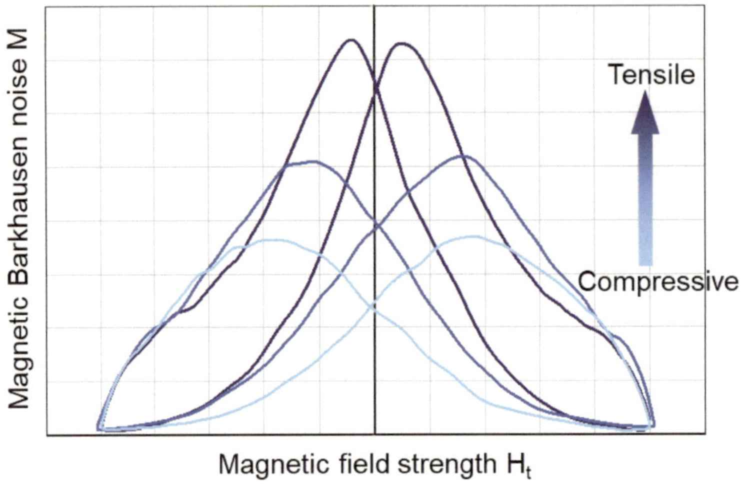

Figure 2.
Stress dependence of the magnetic Barkhausen noise profile curve for a single ferromagnetic sample in a tensile/compressive test in elastic strain range, without a change in the mechanical hardness (after [9]).

mechanical stresses (see **Figure 2**), which makes it suitable for the nondestructive determination of residual stresses.

2.1.4 Micromagnetic multiparameter microstructure and stress analysis (3MA)

Micromagnetic multiparameter microstructure and stress analysis (3MA), developed at the Fraunhofer IZFP, is a solution for the quantitative determination of quality-deciding target parameters such as stress as well as hardness, hardening depth (CHD, SHD or NHD), ductile-to-brittle transition temperature (DBTT), yield and ultimate strength. 3MA uses a combination of several micromagnetic measuring methods such as upper harmonic analysis, Barkhausen noise analysis, incremental permeability analysis and eddy current (EC) impedance analysis [10]. The reason for using more than one measuring method for material characterization is the increased robustness against disturbing influences such as material inhomogeneity or surface treatment (similar to the benefits of having different human senses). The 3MA device generates the drive voltages for the magnetization coil and the eddy current coil in the sensor. The 3MA sensor includes an electromagnet, which excites an alternating magnetic field in the ferromagnetic sample, which, in turn, shows micromagnetic effects (see Section 2.1.2). In the most widespread implementation of 3MA, these effects are registered with a Hall probe and coils included in the sensor as well. Other implementations evaluate the current and voltage at the exciter coil. After preamplification, the measured signals are processed in the 3MA device and a controlling PC. This results in a kind of "magnetic fingerprint" which is used for quantitative and qualitative material characterization. For mechanical-technological material characterization, 3MA has to be calibrated on a well-defined calibration set of samples (with known target properties, e.g., stress or hardness).

2.2 Ultrasonic techniques

Among nondestructive methods, ultrasonic methods are the most and widely used ones. Reasons for this are the applicability to almost all materials and the wide range of wall thicknesses that can be tested. In order to excite ultrasonic waves, two main principles are used. Conventional ultrasonic testing is based on the use of the inverse piezoelectric effect: an electrical voltage is applied to a solid (in this case usually ceramics) which leads to a deformation of the material and an excitation of ultrasonic waves. The (direct) piezoelectric effect is used to receive ultrasonic waves. In order to transfer the ultrasound into the specimen (and to receive it), a couplant is needed. In this regard, mostly water and oil are used. Depending on the situation, this implies disadvantages, such as a contamination of the specimen by the couplant and the limited excitation of wave modes. Nevertheless, piezoelectric ultrasound is widely used as a wide range of frequencies can be generated [11]. Electromagnetically excited ultrasound presents an advantageous alternative to conventional ultrasound testing, especially in rough environments or when special wave modes are to be excited. In principle, an electromagnetic acoustic transducer (EMAT) is consisting of a magnet and a radio frequency coil. The ultrasonic sources are induced in the material surface by alternating electromagnetic fields. Thus, on the one hand, no couplant is needed, and small lift-offs between specimen and transducers are acceptable. On the other hand, electromagnetic ultrasound can only be excited in electrically conductive materials, and the range of frequencies that can be generated is limited [12].

In order to perform measurements, regardless of the excitation mode, transducers are required to send and to receive ultrasonic waves. In this regard, different set-ups are feasible. **Figure 3** shows three typical set-ups that are used to measure the so-called ultrasonic time of flight (TOF), which denotes the time the wave (white) propagates from a transmitter (T) through a medium (green) to a receiver (R). Using reflection mode (see **Figure 3**, middle), generally, a transducer is chosen that serves as transmitter and receiver, and the generated wave is reflected at a medium boundary before it propagates back to the receiver. In addition to the TOF, the amplitude of the transmitted or reflected signal is recorded. If the thickness of the specimen is known, it is also possible to measure the velocity of propagation.

In addition to the ability to detect flaws, cracks, etc., ultrasonic techniques are often used to determine stresses. Reason for this is that all tension and strain changes in materials are subject to the so-called acousto-elastic effect which raises measurable changes regarding the velocity of propagation of ultrasonic waves [13]. Hence, measurements of ultrasonic velocity of propagation are used to detect emerging material and stress changes. Depending on the applied wave modes (longitudinal, transverse (SH, SV)) and their propagation and polarization direction in relation to the stress direction, additional changes of the measured ultrasonic TOF occur (**Figure 4**). The accuracy of evaluated stresses using ultrasonic TOF measurements is mainly determined by two factors, that is, the accuracy of the determined acousto-elastic constants and the accuracy of the measured TOF. According to [14] typical TOF measurements lead to an accuracy in the stress determination of

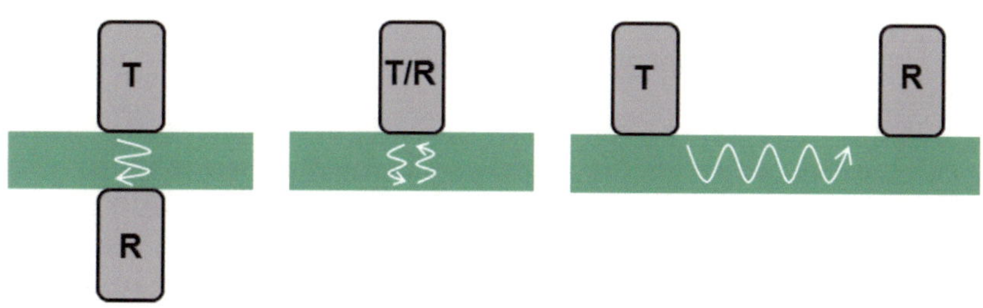

Figure 3.
Three typical set-ups to perform ultrasonic measurements: T denotes the transmitter and R denotes the receiver, and the specimen is marked in green. Left, transmission mode; middle, reflection mode; right, transmission mode.

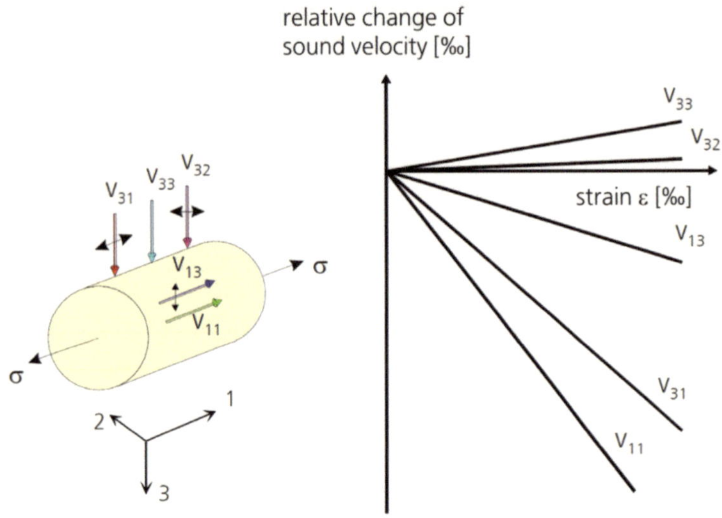

Figure 4.
Relative change in the ultrasound velocity with the elastic tensile strain of a metallic sample: Schematic sketch (after [13]).

approx. ± 20–50 MPa. In order to measure a change of approx. 1% in the TOF, a stress change of approx. 100 MPa is needed. This, however, can only be understood as a rule of thumb as it strongly depends on the used wave mode and the material to be characterized. Measurements of steel rolls showed that longitudinal waves are more sensitive to stress changes than transversal waves: while a change in the stress state of 135 MPa caused a change of 1% in the TOF of the used transversal wave, a change in the stress state of only 87 MPa caused a change of 1% in the TOF of the used longitudinal wave [14].

3. Micro- and macro-residual stresses

Looking into publications accepted by most researchers and engineers, residual stresses are categorized according to the length scale. In this context, a simple classification categorizes residual stresses into macro- and micro-residual stress [15, 16]. Macroscopic residual stresses are stress distributions whose range is several orders of magnitude greater than the microstructure of the material. This type of residual stresses originates, for example, from shot peening, bending, welding and cold hole punching. Microscopic residual stresses, on the other hand, are based on microstructural misfits between phase constituents and grains which can occur as a side effect of phase transformation, intergranular stresses and plastic deformation, for example [17].

According to [2, 15, 18, 19], residual stresses may be distinguished more precisely. The first type or first kind (σ_R^{I}) is the equivalent of macro-residual stresses—the volume average stresses—which has a range of some millimeters. The second type or second kind (σ_R^{II}) analog to micro-residual stresses results from the heterogeneity of grains in polycrystalline materials and is defined as the mean stress value equilibrated inside a grain. The third kind (σ_R^{III}) as (sub)micro-residual stress contains all stresses at any point inside a grain which equilibrate on a scale of several atoms [18–20]. **Figure 5** shows a schematic summary of the three kinds of stresses.

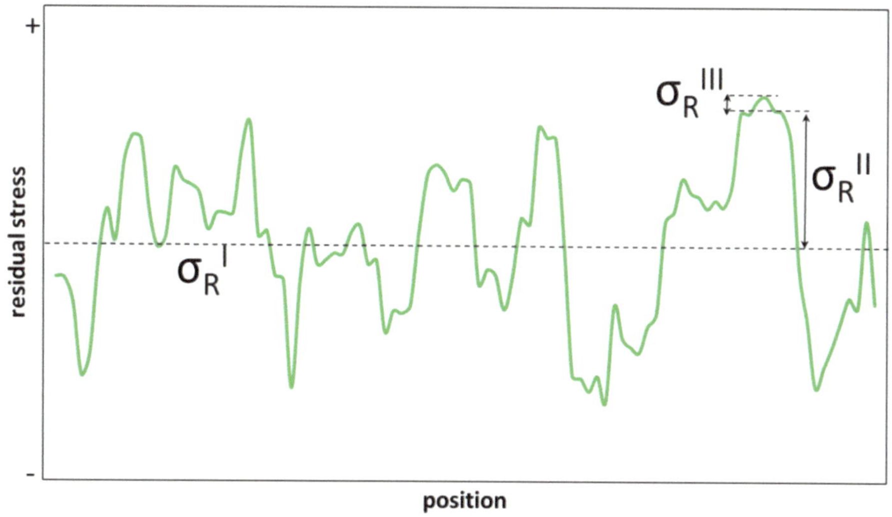

Figure 5.
Three kinds of residual stress over several grains: $\sigma_R{}^I$ represents the average stress over a longer range, that is, across several grains, $\sigma_R{}^{II}$ represents a shorter-range average (i.e., representing one grain or crystal), and $\sigma_R{}^{III}$ represents stress variations within a grain (after [15]).

3.1 Characterization of macro-residual stress

3.1.1 Characterization of macro-residual stress by means of micromagnetic techniques

In the frame of a research project funded by the German Research Foundation (DFG), the residual stress distribution in deep drawn components was examined by means of the micromagnetic method 3MA [9], which is described in Section 2.1.4. The focus of this research project was the examination of high-strength steels in deep drawing, which are increasingly used within the construction of automotive parts due to the reduction of weight with simultaneous enhancement of safety. The reduction of weight is an important contribution to the reduction of primary energy consumption in transportation and so finally adds a contribution to the reduction of CO_2 emissions. However, the application of high-strength steels in deep drawing processes is related to an increase of spring-back. Spring-back is one important criteria for accuracy of the shape geometry. One reason for spring-back is the redistribution of residual stresses when the part is taken from the forming tool [21]. On the other side, the material properties and the deep drawing process parameters are also influencing the spring-back behavior [22]. One aim of the abovementioned project was the determination of the residual stress and its distribution in the deep drawing components with the target to control the process parameters in order to avoid spring-back. The following example demonstrates the correlation between residual stress and spring-back and the potential of the micromagnetic 3MA method for the nondestructive characterization of stress states.

In order to investigate the influence of stress and strain during the deep drawing process on spring-back, six samples with three different rectangular geometries (160 mm × 80 mm), as shown in **Figure 6**, were produced. The three geometries are:

- Rectangular geometry
- Rectangular geometry, trimmed on the small sides of the part in order to remove the influence of the corners on the sample (decreasing stiffness)

- U-shaped geometry

For each geometry, a twin sample was provided.

The 3MA probe was manually and step vice moved along a measuring trace from the middle of the bottom (position 0 according to **Figure** 7) along the wall to the die radius (position 8 according to **Figure** 7). Hereby the samples have been magnetized parallel to the die (or punch) radius by using a magnetization frequency of 250 Hz and a magnetization field strength of 10 A/cm. For each measuring position, 41 measuring parameters derived from upper harmonic analysis, Barkhausen noise analysis, incremental permeability analysis and eddy current impedance analysis were recorded.

For a quantitative determination of residual stress, a calibration procedure is necessary. For the calibration, generally the micromagnetic measuring parameters are correlated to real stress values at a corresponding position, which are measured by a reference method. Bases on these calibration data, a numerical function is generated, which is able to calculate stress values out of the micromagnetic measuring parameters. For the calibration of the shown example, the Barkhausen noise values which yielded the best results out of all the 3MA values were correlated with the tangential component of the residual stress values measured by X-ray diffractometer (XRD). As the samples have been magnetized parallel to the die (or punch), the magnetic measurements are mostly sensitive to the tangential component of the residual stress.

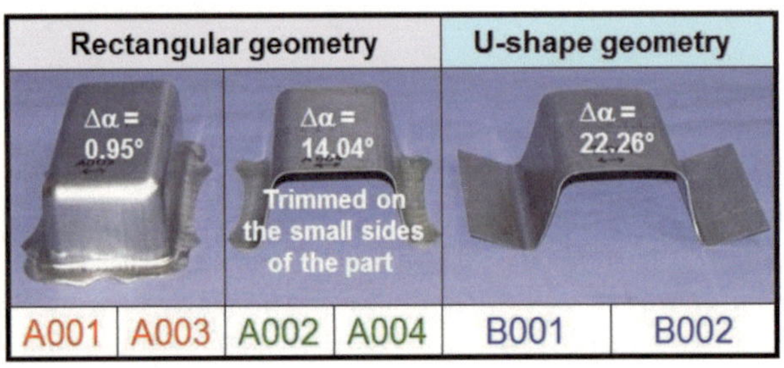

Figure 6.
Examined shapes and correlated spring-back angle Δα (two samples for each shape). The labels below the photos relate to the legend in **Figure** *7 [9].*

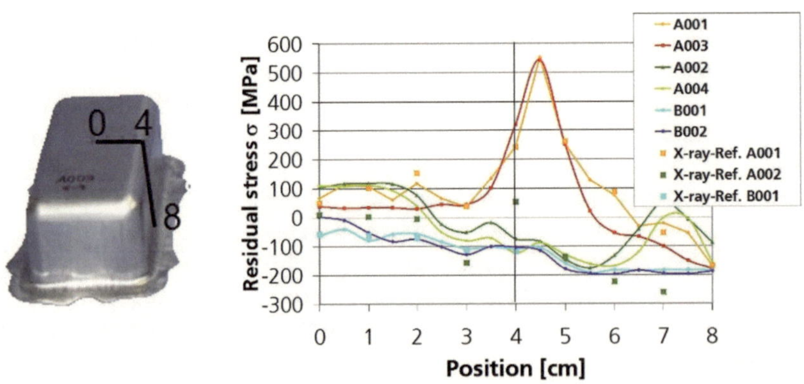

Figure 7.
Residual stress distribution determined micromagnetically by 3MA on a rectangular geometry (A001, A003), a rectangular geometry trimmed on the small sides of the part (A002, A004) and a U-shaped geometry (B001, B002) along a measuring trace from the center of the part bottom (position 0) down to the die radius (position 8) [9].

The graph in **Figure 7** shows the behavior of the micromagnetically determined residual stress on a rectangular geometry, a rectangular geometry trimmed on the small sides of the part and a U-shaped geometry. In the range between positions 4 and 5 cm, a decisive decrease of tensile residual stresses after cutting free is observed. Due to the cutting process and the resulting relief of tensile residual stress, the deep drawn part shows a clear spring-back and hence an increase of the spring-back angle $\Delta\alpha$. The U-shaped sample shows a quite homogenous distribution of stress in a low level along the measured line. Due to the missing sidewall, the sample can expand when taken out from the deep drawing tool, and therefore the spring-back angle increases, and the stress level stays small.

3.1.2 Characterization of macro-residual stress by means of ultrasonic techniques

In this paragraph, a study regarding integral biaxial stress conditions at rolled aluminum sheets for a stretched (0.5%) and an unstretched condition is presented and discussed in order to show the influence of the stretching process on residual stress relief. As for aluminum, the ultrasound velocity is almost equally affected by rolling texture and stress of the material—whereas for steels the ultrasound velocity is affected approximately 10 times more due to the rolling texture compared to stress conditions—these effects need to be separated from each other to determine the existing stress states.

Under the assumption that the stress component along the thickness direction can be neglected due to the texture, the two principal stress directions in and perpendicular to the rolling direction can be determined having information of two directional TOF dependencies. In order to easily implement measurements, an EMAT in impulse-echo mode (cf. **Figure 3** middle) using a center frequency of 2.5 MHz, three bursts and a bandpass filter generating ultrasonic shear waves propagating across the thickness of the sheets, polarized in and perpendicular to the rolling direction by turning the transducer around its own axis at each measuring point, were applied.

Figure 8 shows the distribution of the measured relative TOF changes of the two polarized shear waves due to the two principal stress directions (in and perpendicular to the rolling direction) concerning the formula

$$\textit{Relative}\ \text{TOF}\ [‰] = \frac{\text{TOF in rollingdirection} - \text{TOF perpendicular to the rolling direction}}{\text{TOF perpendicular to the rolling direction}} \times 1000 \tag{1}$$

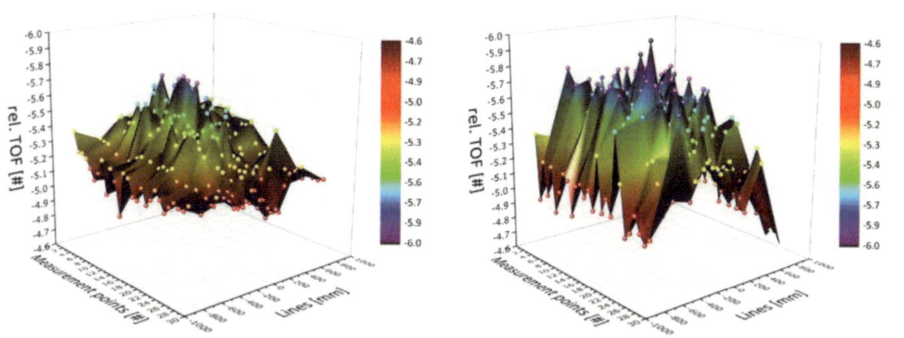

Figure 8.
Distribution of relative TOF measurements at a stretched (0.5%) sheet (left) and an unstretched sheet (right).

at 297 measuring points at a stretched (left) and an unstretched sheet (right) of 30 mm thickness and a flat profile of 2 m × 4 m.

In both diagrams displayed in **Figure 8**, the superimposed influence of the rolling texture and the residual stress state is contained in each measurement point. In order to separate them from each other, the texture needs to be determined by measurements at selected measuring points that have to be cut out in order to relieve the stress condition or at a comparative sample of the same microstructure in a stress-relief heat-treated condition. Afterwards, quantitative stress values can be calculated using the TOF values and the acousto-elastic constants evaluated in tensile tests [13]. Under the assumption of an even distribution of the texture due to the rolling process, the shapes of the measurement results of **Figure 8** reflect the residual stress distribution for the stretched (0.5%) and the unstretched sheet originating from the two principal stress directions.

3.1.2.1 Evaluation of absolute stress in the two principal stress directions for platelike products

For the determination of individual values of two principal stress components at sheets, an additional information concerning the local thickness of the sheets is necessary compared to the approach mentioned above. The principle described in this paragraph can, therefore, be applied using a newly developed hybrid EMAT.

Due to the excitation mechanism, the hybrid EMAT is linked conductively to non-ferromagnetic materials. This kind of EMAT generates two different ultrasonic waves in a time multiplex at the same measurement point. On the one hand, a shear wave propagating across the thickness of the sheet oscillating perpendicular to its direction of propagation is generated, and on the other hand, a longitudinal wave propagating across the thickness oscillating in the same direction is generated.

Depending on the position of the transducer, the anisotropy of the shear wave velocity can be evaluated by turning the transducer around its axis. Since the TOF of the longitudinal wave is not affected by any stress condition across the thickness of the sheet, there is no significant anisotropy for this wave type. Thereafter, reliable information on the thickness of the sheet can be obtained at each measurement point.

Using the evaluated thickness for a stress-free state determined by the longitudinal wave and information on the anisotropy of the two shear waves oscillating in and perpendicular to the rolling direction, the shear wave velocities can be determined. Afterwards, the two main stress states V_{ij} and V_{ik} can be calculated according to Eqs. (3) and (4) below, using the shear wave velocities evaluated previously.

$$\frac{v_{ii} - v_L}{v_L} = \frac{A}{C} \cdot \sigma_i + \frac{B}{C} \cdot (\sigma_j + \sigma_k) \tag{2}$$

$$\frac{v_{ij} - v_T}{v_T} = \frac{D}{K} \cdot \sigma_i + \frac{H}{K} \cdot \sigma_j + \frac{F}{K} \cdot \sigma_k \tag{3}$$

$$\frac{v_{ik} - v_T}{v_T} = \frac{D}{K} \cdot \sigma_i + \frac{F}{K} \cdot \sigma_j + \frac{H}{K} \cdot \sigma_k \tag{4}$$

In Eqs. (2)–(4), σ_i, σ_j and σ_k represent the components of a normalized stress tensor, and v_L and v_T define the ultrasound velocity of a longitudinal (L) and transverse (T) wave for a stress-free condition. v (with its two indices) denotes the ultrasonic velocity for different directions of propagation (first index) and oscillation (second index) for a stress-affected state. A, B, C, D, H, F and K are combinations of the second- and the third-order elastic constants, the quotients of which define weighting factors for the corresponding main stress directions [13].

3.1.2.2 Depth-resolved measurements

Following, an innovative approach is described in order to determine depth-resolved stress conditions using EMATs generating Rayleigh waves in transmission mode in ferromagnetic or conductive materials.

A smart layout of the transducers offers the opportunity to generate and receive Rayleigh waves operating at different frequencies in a time multiplex using the same transducers. The general EMAT concept is explained for ferromagnetic materials here, but it can be applied to non-ferromagnetic ones as well, by modifying the transducer set-up. It is based on the superposition of a perpendicular bias magnetic field B_0 and a high-frequency field generated by a meander-shaped coil that is wound on the ferromagnetic comb structure (**Figure 9**). The generated trace wavelength, λ_s, is defined as the distance of two neighboring wires wound into the grooves of the comb having the same current direction. By winding the coil on just every second or third tooth, respectively, by segmenting the coil into several single coils for each comb tooth using an additionally adequate electrical connecting, the trace wavelength can be varied (λ_{S1}, λ_{S2}, λ_{S3}).

As the penetration depth of a Rayleigh wave is approximately in the range of one wavelength [24], different depth ranges can be achieved. Approaching specific evaluation algorithms comparable to the one mentioned in [25], stress conditions in different depths can be resolved. Since the stress dependence of the ultrasound velocity of a Rayleigh wave is less than for longitudinal and transversal wave types due to the elliptic polarization, the stress resolution is comparatively small.

3.1.2.3 Ultrasonic determination of residual stress and the separation from texture effects on AHSS strips

Advanced high-strength steel (AHSS) enables the automotive industry to increase the stability and crash safety of cars and to reduce weight or CO_2 emissions at the same time. Increasing strip speeds and requirements on the lateral homogeneity of the material properties leads to a need for a development of a nondestructive multisensor solution for the quality assurance. The homogeneity of the texture, grain size and the influence of residual stresses in the material is of great importance for the deformation behavior and decisive for material processing, such as deep drawing or welding. Current in-line NDT systems determine only a subset of the required parameters and do not assess their homogeneity across the strip width.

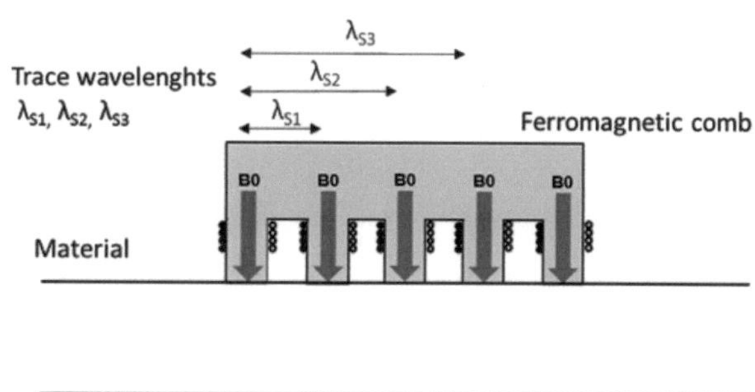

Figure 9.
Set-up of an electromagnetic acoustic Rayleigh wave transducer for generating variable trace wavelengths (after [23]).

In this context, Fraunhofer IZFP developed a hybrid test system for the in-line inspection of AHSS consisting of 3MA and ultrasonic TOF measurements using EMAT.

The challenge in the case of the ultrasonic inspection is the separation of the different effects from texture and residual stresses on the ultrasonic time of flight. The impact on the ultrasonic wave velocity is ranging in a magnitude of percentage to per mil. The objective of the new approach is the use of different wave modes and types to separate the effects. In thin plates, guided waves can be excited with EMAT. By different alignments of the conductive paths to the magnetic field in EMAT, a simultaneous excitation of guided shear horizontal (SH) waves and Lamb waves is possible. Selecting different excitation frequencies and switching between symmetric and antisymmetric modes of different order could be realized.

Because of its symmetry to the texture direction, the lowest symmetric mode of the SH wave (SH_0) could be used to determine the residual stress. The time-of-flight measurement of this wave mode in rolling direction (0°) and perpendicular to it (90°) is therefore independent of the influence of texture. The presence of residual stress is reflected in an asymmetrical behavior in the polar plot (see **Figure 10**). Additionally, the SH_0 mode is dispersion-free and thus independent of thickness, which means that even slight fluctuations in sheet thickness have no influence on the measuring effect.

If the influence of the residual stress is known, the effect can also be calculated for the other wave modes and types, and finally the influence of texture can be determined.

In order to implement the inspection concept, a special arrangement of the probes is required (see **Figure 10**). Thus, the macro-residual stress is determined as an integral value over the sheet thickness and the travel distance of the wave.

3.1.2.4 Residual stress analysis in freight car wheels and research challenges of localized stress evaluation

Residual stress occurring during operation of components can also be detected and quantified with ultrasonic methods. The examination of residual stress in railway wheels of freight cars is an industrial application of ultrasonic stress evaluation that has been established more than 25 years ago [27]. In the production of new wheels, compressive stresses in the tread in the circumferential direction are generated by well-defined quenching, which should prevent the formation of cracks in the wheel-rail contact zone in case of damage to the surface. However, in the case of block-braked freight car wheels, the running surface is subjected to high

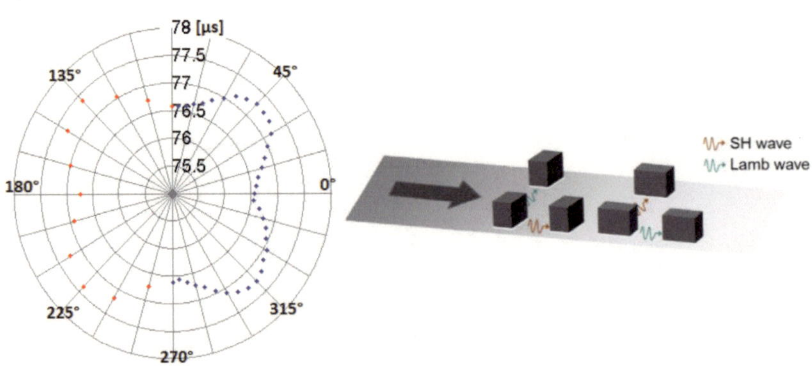

Figure 10.
Time of flight of SH_0 mode depending on the rotation angle, measured on an AHSS plate (left). Arrangement of probes along the steel strip (right, after [26]).

thermomechanical loads, which, depending on the axle load and the route profile, lead to high energy input in the form of heat. In addition, faulty, non-opening brakes cause the material to overheat, which, in turn, results in compressive stress being converted into tensile stress that may increase until rim failure occurs. There are various systems that determine these stress values in the industrial environment by means of ultrasonic methods. The procedures and evaluation steps presented here refer to the only system for the fully automatic determination of circumferential residual stresses. This system represents a development of Fraunhofer IZFP and has been used for more than 25 years in maintenance and wheel production, presently in the fourth generation.

In order to check the wheel for critical stress situations, the acoustic birefringence parallel to the tread is evaluated from the side of the wheel rim. For this task, a linearly polarized shear wave is used, since this wave type exhibits the strongest interaction with the stress in the circumferential direction for the given geometry. The required ultrasonic wave is generated with an EMAT that excites the wave in the ferromagnetic base material of the wheel without a coupling agent (**Figure 11**). Also due to the geometry and surface condition of used wheels, only the relative stress difference with respect to the radial direction of the wheel rim is checked. A measurement of the absolute change in the sound velocity is not possible within the scope of the required accuracy of the sound path determination in the maintenance of the track wheels.

When determining the relative stress difference, the measurement of the ultrasonic time of flight of the linearly polarized wave with its direction of oscillation perpendicular to the running surface is chosen as reference, as the stresses in the radial direction do not change over the life of the wheel according to experience. The test as well as the presentation of the results takes place as a depth profile of the tread in the direction of the wheel web. The typical step size of the measurement is 1 mm. **Figure 11** shows the test positions and the wave propagation of the ultrasound schematically.

Since the acousto-elastic material constant K is known, the measured differences in the ultrasonic time of flight of the two measuring tracks can be converted into a relative stress value between the tangential and the radial direction of the wheel. This is done at each test position according to Eq. (5):

$$\sigma_{tan} - \sigma_{rad} = \mathrm{K}\,\frac{TOF_{tan} - TOF_{rad}}{TOF_{rad}} \tag{5}$$

The result is a stress curve in the depth direction. Each stress value represents an integral result over the entire traveled sound path, as shown in **Figure 12**. Depending on whether a newly manufactured wheel or the review of a used wheel

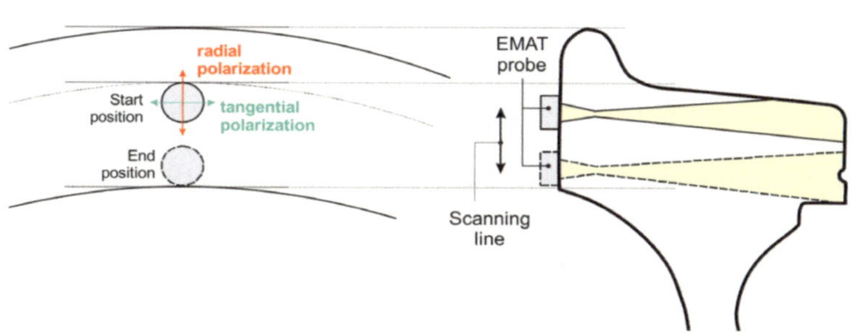

Figure 11.
Principle of the stress evaluation of the railroad wheel. The scan is performed in steps of 1 mm. The EMAT transceiver with the probe generates a linear polarized shear wave.

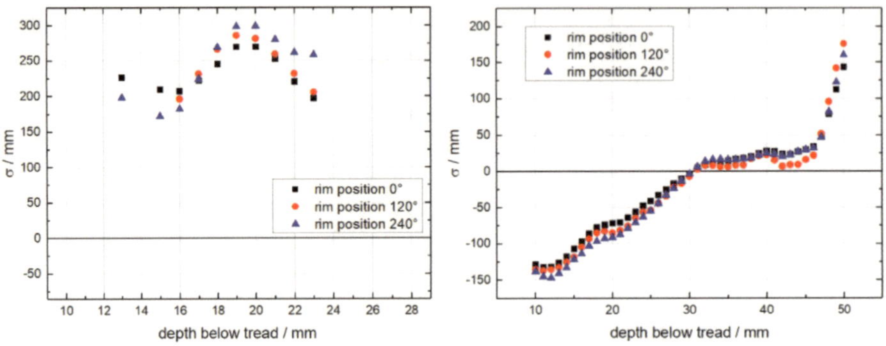

Figure 12.
Typical stress graph of a "R7" wheel in new condition (left) and another wheel of this type in worn and used condition. In each case, tests were performed on each wheel on three positions with an offset by 120°.

is required, different stress limits were applied, which decide on the further operation of the freight car wheel. **Figure 12** shows a comparison between the residual stress of a new and a used wheel.

Now, block-braked track wheels are experiencing a recurrence, also in passenger transport, as a result of which the wheel developers' desire for improved methods for determining local residual stress is emerging. By fundamentally linking finite element computation and knowledge of ultrasonic wave interactions with sample volume stresses, geometrical and thermomechanical states can be predicted. These predictions represent possible solutions for real stress situations leading to the measured integral stresses. Modeling to predict possible wheel residual stress as a first step is a well-known method [28]. The challenges, however, lie in the combination with sufficiently large amounts of residual integral residual stress data as well as the real geometries investigated. With the help of this data, test systems would be able to identify most likely the stress distribution depending on the wheel geometry, which leads to the present measurement result, and thus to make statements about critical local stresses along the sound propagation path.

This procedure for combining preliminary information as a statistical interpretation aid for metrologically inaccessible information beyond the described application is a research field with high potential for improvements in the informational quality of volumetric residual stress evaluation.

3.2 Characterization of micro-residual stress

3.2.1 Characterization of micro-residual stress by means of micromagnetic techniques

3.2.1.1 Volumetric characterization

Micromagnetic testing methods are based on the fact that the ferromagnetic material behavior is influenced by microstructures as well as by applied load or residual stress. Separation of these two influences in micromagnetics can be carried out by analyzing different interaction mechanisms of magnetic structure with microstructure, on the one hand, and with residual or applied stresses, on the other hand. Moreover, it is well-known that microstructure changes always cause changes in the micro-residual stress distribution. This means that, in fact, separating micro-residual stress from macro-residual or load stress indirectly allows for the separation of microstructure from macro-residual stress.

Therefore, a nondestructive micromagnetic testing approach has been developed for the evaluation of material changes independently on the macro-residual stresses [29]. This approach is based on the applied stress dependence of signals obtained from magnetic hysteresis loop at different applied load stresses in the elastic range. The magnetic hysteresis curve changes its shape under influence of applied load stresses depending on whether the applied load stresses are compressive or tensile. It has been observed experimentally that magnetic hysteresis curves recorded at different applied load stresses contain two intersection points that are invariant with changing applied stress. A possible explanation of this effect is that the magnetization processes at these points take place by 180° Bloch wall movements only. 180° Bloch walls have short-range stress fields, whereas 90° Bloch walls have long-range stress fields. As the intersection points do not change their position with changing applied stress, they are stress-independent, and therefore the magnetization processes in these points are based on 180° Bloch wall movements only. Parameters that define these intersection points are the related magnetic field strength H_{sp} and flux density B_{sp} (**Figure 13**). This means that both parameters H_{sp} and B_{sp} characterize the 180° Bloch wall movement only and are affected by short-range residual stresses or microstructure features that can induce them. Microstructure defects, e.g., second-phase precipitates, or dislocations induce such short-range residual stresses.

Furthermore, micromagnetic measurements have been carried out on cylindrical specimens of 20MnMoNi5-5 steel at superimposed elastic compressive and tensile load stress in the range of −150 MPa up to 200 MPa, and the stress invariant points have been determined automatically. In the first step, stress invariant points were determined on stress-relieved specimens of different microstructures with different mechanical properties. X-ray measurements confirmed that the samples were nearly free from residual stress (the maximum stress observed was 50 MPa). It was found that parameter H_{sp} depends almost linearly on the yield strength values of those samples (**Figure 14**). In order to induce microstructure changes, the samples have been plastically deformed to 1 and 2% plastic deformation. After each plastic deformation the samples have been investigated by means of the approach described above. It has been observed that the increase of the plastic deformation leads to an increase of H_{sp}. The change of H_{sp} with increasing plastic deformation depends on the initial mechanical properties of the samples. In the case of the samples with high yield strength, H_{sp} increased almost linearly with the percentage

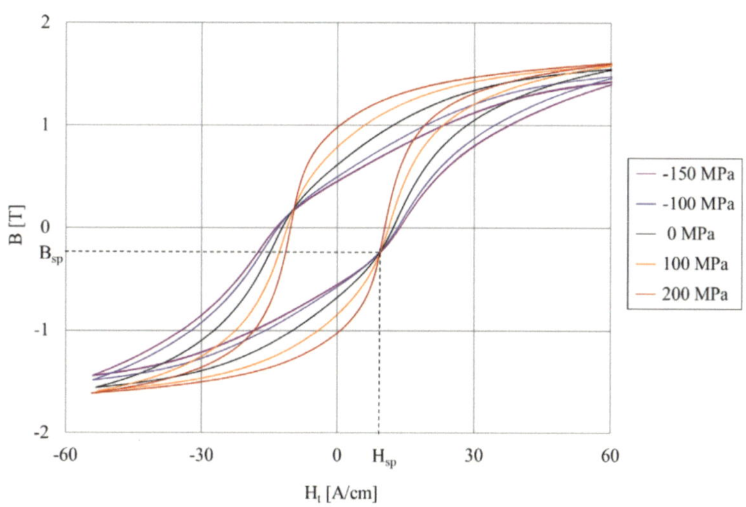

Figure 13.
Hysteresis curves measured as different superimposed load stress levels including the invariant stress points and their corresponding coordinates (after [29]).

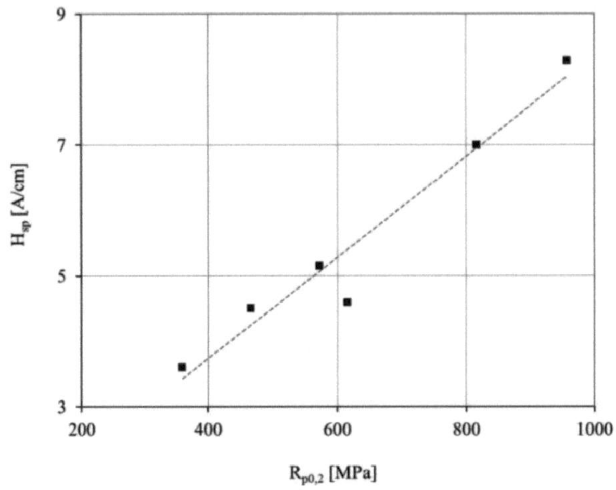

Figure 14.
Dependency of parameter H_{sp} on the yield strength $R_{po,2}$ (after [29]).

of plastic deformation. In the case of the sample with low yield strength, H_{sp} does not change, whereas in the case of the samples with average yield strength, H_{sp} changed up to 1% of plastic deformation and after that remained constant. In the case of soft materials, the dislocation density changes only after higher degrees of plastic deformation. The more mechanically soft a material is, the more malleable it is, due to the absence of microstructure defects. Due to this fact, in the case of mechanically soft microstructures, H_{sp} remains constant for plastic deformation degrees ≤2% and would probably increase at higher plastic deformation degrees. With increasing yield strength (mechanical hardness), the ductility of the materials decreases, and the dislocation density already takes place at smaller plastic deformation degrees. For this reason, H_{sp} increases in the case of mechanically hard microstructures even after a small plastic deformation (<1%).

Moreover, experiments have shown that for samples containing compressive residual stress, stress invariant points occur only when applying compressive stress, whereas for samples containing tensile residual stress, stress invariant points occur when applying either compressive or tensile stress. This means that the sign of applied stress (compressive or tensile) under which the stress invariant point occurs gives information on the sign of the residual stress in the sample.

In practice, this approach is applicable in the context of a monitoring concept if the operating conditions provide for or permit the variation of the operating pressure. A further possible application of this approach is the characterization of the microstructure in the frame of material development, for example, on a material containing macro tensile stress as a result of processing.

The procedure described above allows for the characterization of micro-residual stress induced by dislocations. Micromagnetic measurement techniques based on the tensile load-dependent maximum of the Barkhausen noise amplitude can be used for the analysis of micro (third kind)-residual stress. Such stresses can be induced, e.g., by nanoscale second-phase precipitates due to the difference in lattice parameters of Cu and Fe. For this purpose, Fe-Cu alloys with well-defined amounts of nanoscale Cu precipitates have been manufactured and investigated. The micromagnetic concept is based on load stress-dependent Barkhausen noise measurements. The maximum of the Barkhausen noise amplitude (M_{MAX}) obtained during one hysteresis cycle at a time is recorded under varied tensile load stresses, leading to an $M_{MAX}(\sigma)$ curve that has a relative maximum. A shift of this relative maximum along the stress axis can be observed as a measure for the change of the micro- or macro-residual stress condition. A measurement technique based on this effect permits the quantitative characterization of residual stress variations without the use of a reference method such as X-ray diffraction. If the superimposed residual stress is of the tensile type, the Barkhausen noise activity of the iron-based materials is more enhanced than in the stress-free condition, and the curves reach their maximum at lower load stresses, that is, the curve shifts to the left-hand side and in the other direction in the case of superimposed compressive stress. In order to determine the micro-residual stresses induced by nanoscale Cu particles only (i.e., to eliminate the influence of macro-residual stress from quenching and micro-residual stress from different thermal expansion coefficients of particles and surrounding matrix), micromagnetic measurements have been performed in three steps, which are described in [30, 31]. This procedure permits the volumetric evaluation of micro-residual stresses.

3.2.1.2 Imaging of local residual stress distribution using Barkhausen noise and eddy current microscopy (BEMI)

The fundamental sensor principle of 3MA has been implemented in many technological variants and sizes, even down to where microscopic lateral resolution is reached: The Barkhausen noise and eddy current microscope (BEMI) is a scanning probe device based on miniature micromagnetic sensors [32]. So far, two sensor designs have been implemented as BEMI modules:

- *Ferrite core sensor*: Copper coils wound on a small ferritic core with a 300-nm-wide gap facing the sample surface are used for picking up magnetic Barkhausen noise (MBN) during cyclic magnetization of the sample, for characterizing permeability and conductivity in an eddy current (EC) operation mode and for measuring eddy current incremental permeability (ECIP). The MBN, EC and ECIP modes are performed alternatingly using a 3MA controller device. Depending on the material, a lateral resolution of about 10–20 μm is achieved [33].

- *Point probe:* A needle-shaped ferromagnetic core equipped with a primary (excitation) and a secondary (detection) coil is applied for characterizing local permeability and magnetic field [34]. The sample properties close to the needle tip strongly affect the secondary coil voltage waveform under application of sinusoidal voltage to the primary coil. Point probes deliver isotropic lateral response at a resolution depending on the sharpness of the needle tip. Experiments have proven that some features can be determined with a resolution of around 20 μm [35].

For both sensor types, an evaluation software determines the characteristic features from the signals received. As further explained in the previous sections of this contribution, magnetic Barkhausen noise amplitude and magnetic permeability are strongly modulated by local stress for a given material. This leads to a lateral stress-dependent contrast in feature images produced by BEMI. Multiple mathematical and procedural approaches are applied for predicting stress quantitatively in an indirect way, using these micromagnetic features as input.

On the example of locally laser-treated X20Cr13 steel, the stress sensitivity of features obtained with both probe types is demonstrated. The sample shown in **Figure 15(a)** was scanned with a point probe, **Figure 15(b)**, and ferrite core sensor, **Figure 15(c)**. The results, as shown in **Figure 16**, indicate strong stress fields surrounding the laser-treated spots. The actual stress values were determined with

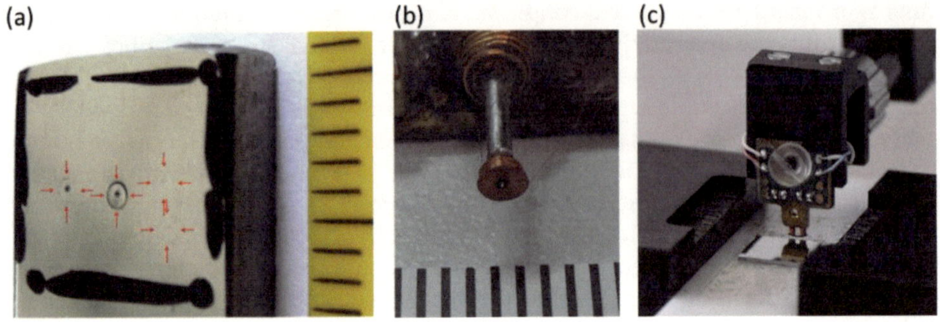

Figure 15.
(a) Photo of X20Cr13 steel sample with laser-heated spots (scale unit is millimeters) [35]. (b) Photo of point probe [35]; (c) photo of ferrite core sensor (Fraunhofer IZFP).

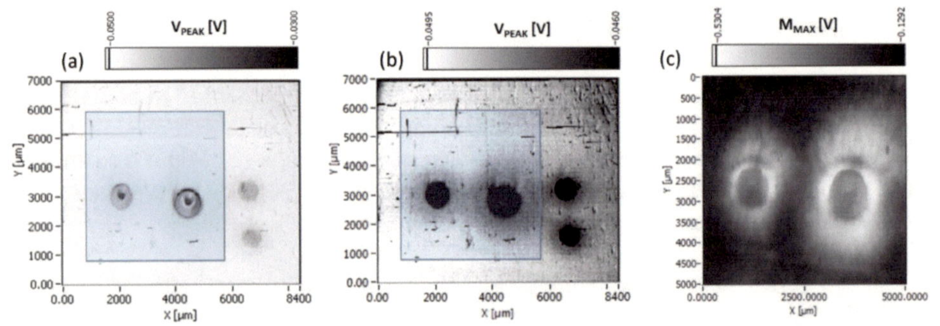

Figure 16.
(a) Scan of point probe feature V_{PEAK} in the area marked in **Figure 15(a)**; *(b) the same V_{PEAK} scan with narrow grey scale value range exposing stress fields around the laser-treated spots; (c) magnetic Barkhausen noise peak M_{MAX} of two laser-hardened spots inside the framed area [33, 35].*

X-ray diffraction and indicated peak stresses M ranging from −300 MPa up to +600 MPa.

The novel micro-residual stress mapping method based on magnetic Barkhausen noise (RESTMAB) approach allows for a nondestructive estimation of micro-residual stress of ferromagnetic materials without any help from conventional stress measurement methods. Moreover, RESTMAB is a timesaving approach for quantitative stress determination with BEMI.

As previously mentioned, the position of the maximum of the $M_{MAX}(\sigma)$ curve is proportional to micro-residual stress (**Figure 17(a)**). As an example, the micro-residual stress inside a pre-deformed sample is calculated using Eq. (6):

$$\sigma_R = \sigma_{crit.(n)} - \sigma_{crit.(p)} \quad (6)$$

where σ_R is the residual stress and $\sigma_{crit.(n)}$ and $\sigma_{crit.(p)}$ are the applied stresses in the maximum of the $M_{MAX}(\sigma)$ curve for the non- and pre-deformed samples, respectively.

RESTMAB works based on the information extracted from the $M_{MAX}(\sigma)$ curve and using regression analysis for calibration. The residual stress and the M_{MAX} values are the inputs of the regression analysis. However, the $M_{MAX}(\sigma)$ curves require an adjustment prior to calibration. This adjustment is required because the MBN response is related to the sum of applied and residual stress. In other words, the MBN output shows the effect of residual stress and applied stress at the same time. Therefore, the M_{MAX} values in **Figure 17(a)** are affected by the residual and applied stress, while the stress axis just presents the applied stress. As the separation of the effect of residual and applied stress for MBN is not possible, the stress axis is adjusted to show the sum of applied and residual stress. This is achieved by horizontally aligning the curves in their respective maxima. As shown in **Figure 17(b)**, the $M_{MAX}(\sigma)$ curves of the pre-deformed samples were moved to the $M_{MAX}(\sigma)$ curve of the non-deformed sample in such a way that all the maxima were aligned under the maximum of the $M_{MAX}(\sigma)$ curve of the non-deformed sample [20]. The residual stress σ_R of each sample is then measured based on the peak shift compared to this reference, using Eq. (7):

$$\sigma_{tot.} = \sigma_{app.} + \sigma_R \tag{7}$$

where $\sigma_{tot.}$, $\sigma_{app.}$, and σ_R are total, applied, and residual stresses, respectively. After modification of the stress axis, the modified stress and M_{MAX} parameters are used as input for a regression analysis in order to determine a polynomial expression for the relation between MBN parameters and (total) stress.

Figure 18 shows the $M_{MAX}(\sigma)$ curves before and after adjustment for non-deformed, 1 and 3% pre-deformed high-strength steel, respectively, and **Figure 19** shows the relation between measured and estimated stress after calibration. The measured stress values are obtained by the summation of the applied stress determined by the load cell of the tensile machine and the residual stress of the sample calculated by means of the shift between the peak positions at the $M_{MAX}(\sigma)$ curves (**Figure 18**). The estimated stress values show the residual stress calculated by means of the RESTMAB procedure.

Determination of the residual stress distribution in an arbitrary area with high lateral resolution is possible using BEMI calibrated with RESTMAB (**Figure 20**)

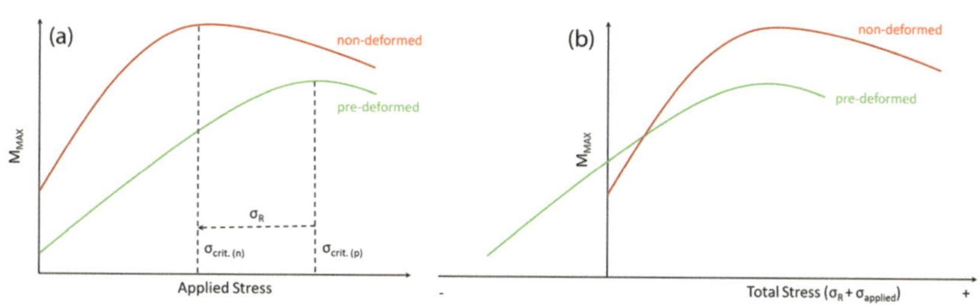

Figure 17.
(a) Schematic $M_{MAX}(\sigma)$ curves for non- and pre-deformed material; (b) curves after horizontal alignment in the respective maxima [20].

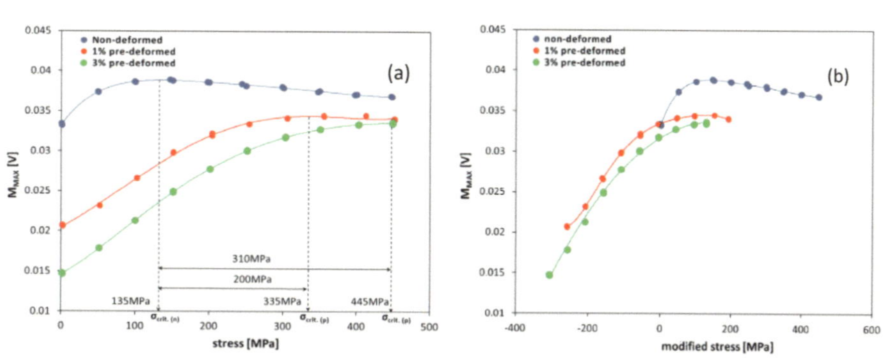

Figure 18.
(a) Measured $M_{MAX}(\sigma)$ curves; (b) adjusted $M_{MAX}(\sigma)$ curves for the calibration process [20].

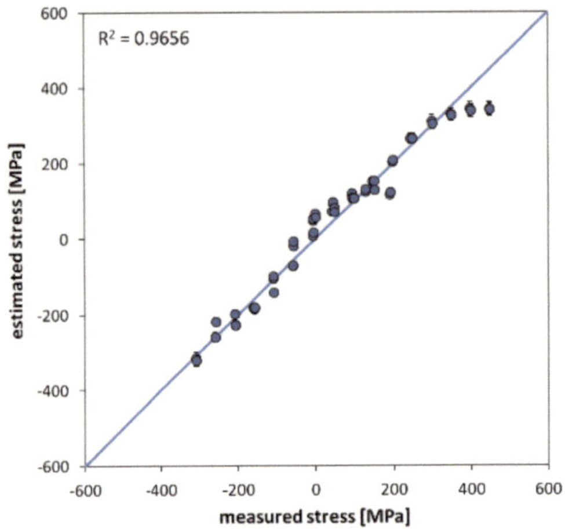

Figure 19.
Relation between measured residual stress and estimated residual stress with RESTMAB for high-strength steel [20].

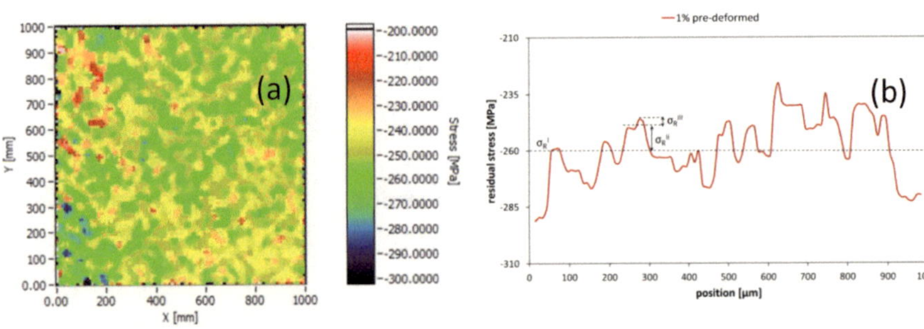

Figure 20.
(a) Micro-residual stress distributions measured with the calibrated BEMI using the RESTMAB for 1% pre-deformed sample. (b) Line scan extracted from the scan, in which it shows different orders of residual stresses measured with the calibrated BEMI for high-strength steel [20].

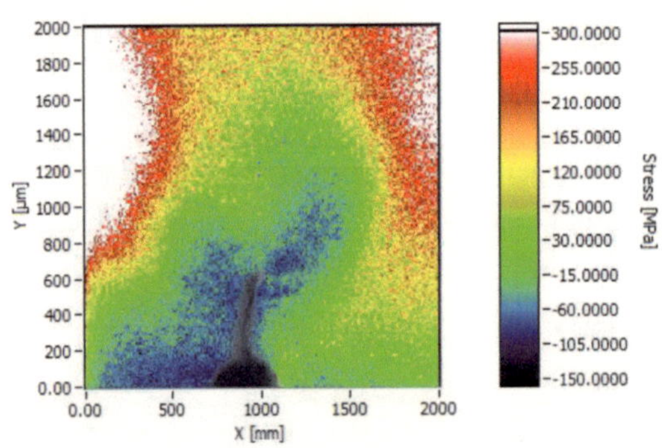

Figure 21.
Residual stress distribution in front of a crack tip in high-strength steel using BEMI calibrated by means of RESTMAB. The overlaying sketch in dark gray shows the notch and crack [20].

illustrates the calibrated BEMI scan of pre-deformed sample of the high-strength steel, in which it shows the stress distributions of the sample in the scan area.

In another application, BEMI calibrated with RESTMAB was used to determine the stress distribution in front of a fatigue crack tip. **Figure 21** illustrates the stress distribution as well as the plastic zone in front of a crack tip. Similar BEMI scans of a crack tip were reported by Altpeter et al. [32]; however, in their approach, BEMI was calibrated using XRD data.

3.2.2 Characterization of micro-residual stress by means of ultrasonic techniques

Residual stresses on the microscale require ultrasonic methods, which are characterized by a low interaction volume and high frequencies. A common method in research is scanning acoustic microscopy (SAM), which uses high-frequency ultrasonic waves >100 MHz to achieve lateral resolution in the range of 50–100 microns. By increasing the frequencies up to 1 GHz, the resolution is further improved. Surface waves or near-surface waves excited by such microscopes are used to determine the sound velocity and to map two-dimensional surface stress states [36]. To determine the absolute stress values, the sound velocity for the phase of the material to be examined must first be determined in calibration experiments, usually in tensile tests with parallel ultrasonic measurement. Since this can be a problem especially with polycrystalline materials due to local orientation differences, the method is also often used to represent only local anisotropy.

Based on the range of materials that can be examined, the use of a SAM is determined solely by the sound-damping properties [37]. Thus, therefore polymers and ceramic materials can be examined. Due to the small working distance and the sensitive sapphire or fused quartz lenses, a high surface quality and thus preparation are necessary.

Although the method has meanwhile lost in importance, current material developments in the field of functionalized and smart materials pose new challenges for the test methods that are used to describe their condition. Here, the SAM could be a useful extension, for example, to investigate non-metallic materials that are functionalized by the introduction of microscopic residual stress fields.

4. Conclusions

In this contribution, nondestructive methods are discussed with respect to their ability to characterize residual stresses. It is shown that there are approaches and testing situations where both ultrasonic and micromagnetic techniques are able to (at least quantitatively) determine residual stress. Nevertheless, the contribution clarifies that ultrasonic techniques are mainly used to determine macro-residual stress, whereas micromagnetic techniques are mainly used to determine micro-residual stresses. Most presented techniques, such as BEMI, 3MA, different EMAT applications, etc., are available for industrial use. New challenges can be found especially in the context of a depth-resolved determination of residual stresses.

Acknowledgements

The authors would like to thank the German Research Society for the financial support of research activities under grants Al 442/5-2, SCHM 746/71-2 and KI 1135/2-2.

Moreover, the authors would like to thank the Federal Ministry for Economic Affairs and Energy for the financial support of research activities under grant 1501523.

Notes/thanks/other declarations

The authors would like to thank Dr. Farajian for the performance of X-ray measurements on specimens made of 20MnMoNi5-5 steel.

Scientific innovation and technological progress can only be achieved when motivated people work together in a fruitful atmosphere. Only if these conditions are fulfilled and only if appropriate funding is provided, new technologies can be developed. Thus, the authors of this paper want to thank all funders and investors, all industrial and academic project partners as well as everyone who contributed in any way to the presented results.

Author details

Madalina Rabung*, Meisam Amiri, Michael Becker, Melanie Kopp, Ralf Tschuncky, Ines Veile, Fabian Weber, Miriam Weikert-Müller and Klaus Szielasko
Fraunhofer Institute for Nondestructive Testing IZFP, Saarbrücken, Germany

*Address all correspondence to: madalina.rabung@izfp.fraunhofer.de

References

[1] Hosford W. Mechanical Behavior of Materials (2nd ed). New York: Cambridge University Press; 2010. p. 410

[2] Macherauch E. Introduction to residual stress. In: Niku-Lari A, editor. Advances in Surface Treatments Vol. IV - Residual stresses. Oxford, New York, Frankfurt: Pergamon Press; 1987. pp. 1-36

[3] Weiss P. L'hypothese du champ moléculaire et de la proprieté ferromagnétique. Journal de Physique. 1907;**6**:661-690

[4] Kneller E. Ferromagnetismus. Berlin, Heidelberg: Springer; 1962

[5] Cullity BD. Introduction to Magnetic Materials. Boston: Addison-Wesley. ISBN: 0-201-01218-9; 1972

[6] Tschuncky R. Sensor- und geräteunabhängige Kalibrierung elektromagnetischer zerstörungsfreier Prüfverfahren zur praxisorientierten Werkstoffcharakterisierung. PhD thesis (dissertation). Saarbrücken, Germany: Saarland University; 2011

[7] Seeger A. Moderne Probleme der Metallphysik. Berlin, Heidelberg, New York: Springer; 1966

[8] Barkhausen H. Zwei mit Hilfe der neuen Verstärker entdeckte Erscheinungen. Physikalische Zeitschrift. 1919;**20**:401-403

[9] Altpeter I, Sklarzcyk C, Kopp M, Kröning M, Hübner S, Behrens BA. Nondestructive characterizing stress states in conventional deep drawing processes by means of electromagnetic methods. In: Shin Y-K et al., editors. International Workshop on Electromagnetic Nondestructive Evaluation. Amsterdam, Washington, Tokyo: IOS Press; 2009. pp. 131-139

[10] Altpeter I, Becker R, Dobmann G, Kern R, Theiner WA, Yashan A. Robust solutions of inverse problems in electromagnetic non-destructive evaluation. Inverse Problems. 2002;**18**: 1907-1921

[11] Krautkrämer H, Krautkrämer J. Werkstoffprüfung mit Ultraschall. Berlin, Heidelberg, New York: Springer; 2013 (Nachdruck von 1986)

[12] Niese F. EMUS-Wanddickensensor für die Pipeline-Inspektion mit integrierter Wirbelstrom- und Streuflussprüfung [Ph.D. thesis]. Saarbrücken: Universität des Saarlandes; 2010

[13] Schneider E. Untersuchung der materialspezifischen Einflüsse und verfahrenstechnische Entwicklungen der Ultraschallverfahren zur Spannungsanalyse an Bauteilen [Ph.D. thesis]. Stuttgart: Fraunhofer IRB Verlag; 2000; ISBN 3-8167-5563-1

[14] Schneider E, Herzer HR, Hübschen G, Wildau M, Steinhoff K. Bestimmung des oberflächennahen Spannungszustandes von Walzen in DGZfP-Jahrestagung 2009. ZfP in Forschung, Entwicklung und Anwendung (2009). Deutsche Gesellschaft für zerstörungsfreie Prüfung (DGZfP); 2009

[15] Withers PJ, Bhadeshia HKDH. Residual stress: part 1 – measurement techniques. Materials Science and Technology. 2001;**17**:355-365. DOI: 10.1179/026708301101509980

[16] Hauk V. Structural and Residual Stress Analysis by Nondestructive Methods. Amsterdam: Elsevier; 1997. DOI: 10.1016/b978-0-444-82476-9.x5000-2

[17] Withers PJ, Bhadeshia HKDH. Residual stress: Part 2 – Nature and

origins. Materials Science and Technology. 2001;**17**:366-375. DOI: 10.1179/026708301101510087

[18] Kloos KH. Eigenspannungen, efinition und Entstehungsursachen. Zeitschrift Werkstofftechnik. 1997;**10**: 293-302. DOI: 10.1002/mawe.19790100906

[19] Totten G, Howes M, Inoue T, editors. Handbook of Residual Stress and Deformation of Steel. Ohio: ASM International; 2002

[20] Amiri M. On the Microstructural Stress and Strain Behavior Determination of Metallic Materials Based on Electromagnetic Phenomena [Ph.D. Thesis]. Saarbrücken: Saarland University; 2015

[21] Hütte H. Rückfederungsverhalten hochfester Stahlblechwerkstoffe beim Tiefziehen [PhD thesis]. Universität Hannover; 2001

[22] Doege E, Hütte H, Elend LE. 2nd Int. Conf. and Exhibition on Design and Prod. Of Dies and Molds. Turkey; 2001

[23] Salzburger H J, Höller P. Ultraschallprüfung an Komponenten bei Temperaturen bis 300°C. 14. MPA-Seminar. Titel Hauptwerk: 14. MPA-Seminar '88. Sicherheit und Verfügbarkeit in der Anlagentechnik mit dem Schwerpunkt Langzeitintegrität der druckführenden Bauteile von Kernkraftwerken; 1988

[24] Viktorov IA. Rayleigh and Lamb Waves, Physical Theory and Applications. New York: Plenum Press; 1967

[25] Mora P, Spies M. In: Inverse Problems: Inversion of Residual Stress Profiles from Ultrasonic Rayleigh Wave Dispersion Data, Bristol: IOP Publishing; 34,055001, 2018

[26] Weikert M, Niese F, Szielasko K. Neue Ansätze zur Charakterisierung hochfester Stähle mittels geführter Ultraschallwellen. in DGZfP-Jahrestagung 2014. ZfP in Forschung, Entwicklung und Anwendung (2014): Deutsche Gesellschaft für zerstörungsfreie Prüfung (DGZfP). Mi.2.B.4; 2014

[27] Schneider E et al. Reliability assurance of railroad wheels by ultrasonic stress analysis. In: European Conference on Residual Stresses. Oberursel: DGM-Informationsgesellschaft; 1993. pp. 441-450. ISBN: 3-88355-192-9

[28] Residual Stresses in Railroad Commuter Car Wheels; Research Result. US Department of Transportation-Federal Railroad Administration. RR99–01; April 1999

[29] Tschuncky R, Rabung M, Youssef S, Farajian M, Luke M. Separation of residual stress and microstructure influences in nondestructive micromagnetic materials characterization for evaluation of power plant safety. Final report of Grant No. 1501523 in the frame of German national program for reactor safety; 2017

[30] Altpeter I, Dobmann G, Kröning M, Rabung M, Szielasko K. Micro-magnetic evaluation of micro residual stresses of the IInd and IIIrd order. NDT&E International. 2009;**42**:283-290. DOI: 10.1016/j.ndteint.2008.11.007

[31] Rabung M, Altpeter I, Boller C, Dobmann G, Herrmann HG. Non-destructive evaluation of the micro residual stresses of IIIrd order by using micro magnetic methods. NDT&E International. 2014;**63**:7-10. DOI: 10.1016/j.ndteint.2013.12.011

[32] Altpeter I, Bender J, Hoffmann J, Kopp M. Barkhausen noise and eddy current microscopy, a new scanning probe technique for microscale characterization of materials. Praktische Metallographie. 1998;**35**(3):126-135

[33] Szielasko K, Lugin S, Kopp M, Altpeter I. Barkhausen noise and eddy current microscopy - a new scanning probe technique for microscale characterization of materials. In: Proc. of SPIE. Vol. 5392. Bellingham, WA; 2004. pp. 105-113

[34] Amiri M, Szielasko K, Altpeter I, Boller C, Fischer D, Hahn J. Point probes: A new generation of magnetic sensors for the measurement of local magnetic fields. In: Proc. of the 6th EEIGM Intl Conf on Adv Materials Research, IOP Conf Sereies 31. 2012. p. 012021. DOI: 10.1088/1757-899X/31/1/012021

[35] Augenstein E. Mikromagnetische Werkstoffcharakterisierung mit Punktsonden [thesis]. Germany: Fachhochschule Osnabrück; 2010

[36] Meeks SW et al. Microscopic imaging of residual stress using a scanning phase-measuring acoustic microscope. Applied Physics Letters. 1989;**55**:1835. DOI: 10.1063/1.102326

[37] Lee YC, Kim JO, Achenbach JD. Measurement of stresses by line-focus acoustic microscopy. Ultrasonics. 1994;**32**(5):359-365. DOI: 10.1016/0041-624X(94)90105-8

Chapter 4

Recent Advancements in the Hole-Drilling Strain-Gage Method for Determining Residual Stresses

Emilio Valentini, Lorenzo Bertelli, Alessio Benincasa and Simone Gulisano

Abstract

The hole-drilling Strain-Gage method is a widely used and cost-effective technique for the evaluation of residual stresses. The test method is standardized by ASTM E837-13a, which defines the scope, measurement range, minimum requirements of instrumentation, test procedure, and algorithms and coefficients for the computation of uniform and non-uniform stress distribution. However, the standardized test method presents some limitations regarding the scope and measurement range; moreover, some typical errors involved in the measurements are not taken into account, i.e., errors due to the hole eccentricity, the local plasticity, the intermediate thickness, and the hole-bottom chamfer, which can affect the results in some cases. Also, the standard does not provide the user with a complete guide regarding the evaluation of the uncertainty connected with this type of measurement. The paper presents a more general approach that allows the correction of some errors and overcomes and some limitations of the ASTM E837-13a test method, contributing to greater accuracy of the test results.

Keywords: residual stress, hole-drilling method, ASTM E837, eccentricity error, local plasticity effect, intermediate thickness, hole-bottom chamfer, uncertainty evaluation

1. Introduction

The hole-drilling method is one of the most cost-effective and simple methods to evaluate the residual stresses present in typical industrial workpieces. Those stresses are induced in the material whenever it is subjected to mechanical or thermal treatments and their effect is often a requirement for the best working condition of the workpiece. Therefore, in most industrial applications, it is very important to have an accurate estimation of the magnitude of the residual stresses particularly where they could represent a critical aspect for the integrity of a component. Nowadays, different methods can be adopted in order to measure the residual stresses in a specimen, and they differ in many features, such as, the depth of investigation or the type of material.

The hole-drilling method represents an interesting solution for the measurement of residual stress. Following the instructions reported in the ASTM E837-13a standard [1], a small hole is made in the center of a strain gage composed of a minimum of three grids. The method requires a sequence of drilling steps to be performed, at the end of which the relaxed strains are acquired from the rosette. The data are then used to calculate the magnitude and the direction of the residual stresses using the calibration coefficients supplied by the standard.

The field of application of the method is wide-ranging and comprises typical mechanical engineering sectors, such as, metallurgy, automotive, aerospace and energy. Hole-drilling measurements can be performed in metal, composite and polymer materials [2]. The method also allows the test to be performed in different conditions, both in a laboratory and on field [3, 4].

According to the ASTM standard, the acquired strains are used for the calculation of residual stress applying the appropriate calibration coefficients depending on the thickness of the workpiece and the type of calculation (uniform or non-uniform). In the case of non-uniform distribution, the evaluation of residual stresses is based on the application of the integral method introduced by Schajer [5, 6].

Standard ASTM E837-13a reports the calibration coefficients for three types of standardized rosette (type A, type B and type C), specifying the related geometric dimensions. Unfortunately, the geometric dimensions of many commercial rosettes do not match those of standard rosettes, and in these cases, the calibration coefficients must be recalculated, taking into account the actual dimensions.

Beghini et al. [8, 9] introduced the influence function approach for a blind hole in a thick workpiece. The strain field was computed starting from a database of numerical solutions, implementing a specific geometric configuration in which the components of eccentricity are merely introduced as the geometry parameters rather than being considered as a source of error. This approach is more extensive with respect to the integral method and includes a parametric description of the strain gage rosette. Using the influence function approach, it is possible to include in the calculation different influence parameters such as the thickness of the specimen or a small bottom chamfer that could be present in the drilled hole.

The eccentricity is the possible error that can be made by the operator during the hole-drilling test, due to misalignment between the drilled hole and the strain gage circle. The ASTM standard sets the limit of 0.004 D (where D is the diameter of the gage circle) as the maximum error of eccentricity. If the actual eccentricity value is measured after drilling, the correction of the eccentricity error will increase the accuracy of the measurements. With this correction, the measurements could be acceptable even in cases where the limit set by the standard is exceeded. Over the years, several approaches have been proposed for the correction of this effect [8–14].

Regarding the thickness of the workpiece under testing, the ASTM standard does not include any instructions for residual stress evaluation for intermediate thickness values ranging from 0.2D to D, where D is the diameter of the strain gage circle. Nevertheless, industrial reality presents a lot of cases where this condition exists, such as with metal sheets or automotive parts. To solve this problem, Abraham and Schajer [15] and Beghini et al. [16] presented correction methods with related calibration coefficients.

The calibration coefficients provided by the ASTM E837-13a standard refer to a cylindrical hole with a flat bottom. Sometimes, end mills used in drilling have a small chamfer that produces a hole that is not perfectly cylindrical, with consequent errors in the measurement of relaxed deformations. To obtain measurements that are more accurate, it is therefore necessary to use calibration coefficients that consider the presence of the chamfer at the bottom of the hole. This source of error will be examined in greater detail, considering the proposed calibration coefficients to take this effect into account [17–19].

Sometimes, the hole-drilling method is used on components that have high residual stress values, comparable with the yield stress of the material. Drilling a hole locally modifies the geometry of the specimen and the stresses around the hole increase by the concentration factor of the discontinuity. It is possible to adopt a stress correction methodology for a uniform stress field and the blind hole case [20–22].

Evaluation of measurement-related uncertainties is also analyzed. Although the ASTM E837 standard does not include a detailed method for the uncertainty evaluation, some scientific works define the main sources of uncertainty and propose some approaches for expressing uncertainty in the case of uniform and non-uniform residual stresses [23–27].

Lastly, the paper shows an application of the hole-drilling method on a known-stress testing configuration obtained using a 4-point bending stress condition. The stresses are evaluated taking into account the effect of the correction of some of the errors described above, and then compared with the expected bending stress distribution. Finally, measurement uncertainty is evaluated with the same calculated stress distribution.

2. Improvements in procedure and devices used to perform hole-drilling residual stress measurements

The ASTM E837 standard provides several details about the testing procedure for strain-gage hole-drilling measurements, including the requirements for the entire measurement chain used for performing the test. A typical hole-drilling measurement chain is composed of two parts: the device used for drilling the hole and the strain gage amplifier used for the acquisition of strains.

The drilling sequence is performed using a drilling technology that minimizes the machining-induced residual stresses at the hole boundary. For this reason, the standard requires that drilling speed remains in the range of 20,000–400,000 rpm: this drilling speed can be obtained using either a high-speed air turbine or an electric motor.

The hole can be made using center-hole drilling or the orbital drilling technique. The orbital method has the advantages of adjusting the diameter of the hole by choosing the offset, of determining a more regular flow of chips and of reducing the geometric dimensions of the small chamfer at the bottom edge of the hole (**Figure 1b**).

The ASTM standard defines some features of a hole-drilling device and the fundamental requirement that the drilling depth must be accurately controlled. The uncertainty of the depth increments, required by the standard, depends on the size of the strain gage rosette used during the measurement and needs to be lower than $\pm 0.004D$. The depth accuracy requirement is essential in the case of a non-uniform stress profile where the hole is made using a step-by-step drilling sequence. For example, when using a rosette with a diameter D of approximately 2.56 mm, the depth uncertainty must be lower than ± 10 μm. This requirement can be difficult to obtain using a manual drilling device.

The use of an automatic drilling system instead of a manual system significantly increases the accuracy of the measurements. The automatic systems use an electronic device and dedicated acquisition and control software; they allow accurate control of the positioning of the end mill, necessary to meet the accuracy requirements of the standard and reducing the total testing time (**Figure 2**).

When maximum accuracy is required in depth increments, it is advisable to use an LVDT sensor, connected to the mechanical body of the drilling unit, which

Figure 1.
(a) LVDT sensor installed on the drilling device and (b) orbital drilling slide.

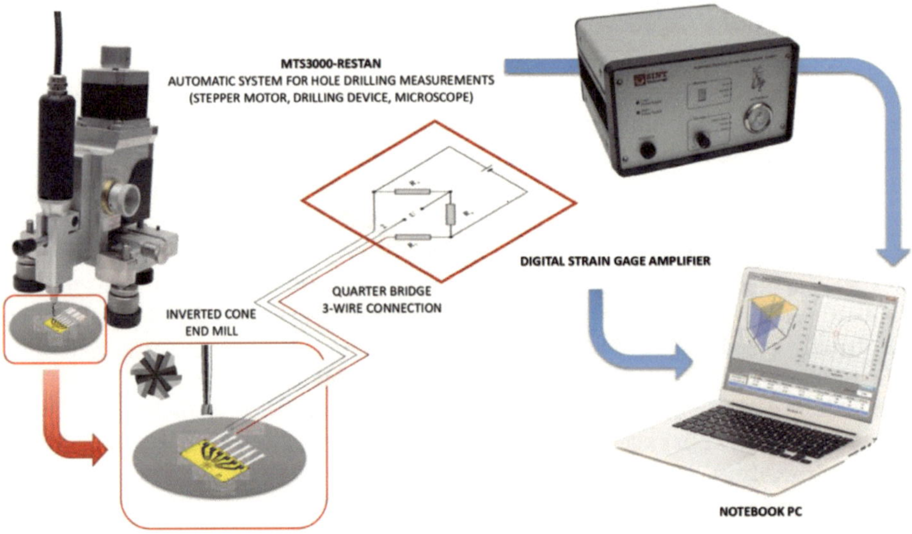

Figure 2.
Hole drilling measurements: typical automatic measuring device (MTS3000-Restan system—SINT Technology).

allows an accuracy of few microns (**Figure 1a**). An automatic system allows a higher number of drilling steps, a uniform feed rate and a fixed stabilization time. Moreover, it can be controlled remotely, minimizing operator presence near the drilling unit [2]; this is particularly important in the case of hole-drilling measurements on polymeric or composite materials.

To obtain accurate measurements, it is very important to establish the point that corresponds to the "zero" cutter depth. The standard identifies it as the point at which the end mill begins to lightly scratch the surface of the workpiece, during slow advance drilling. It is clear that the quality of the results of this process greatly depends on the skill of the operator who carries out the measurements and may not be very accurate.

In the case of tests on conductive materials, it is possible to use the electrical contact technique that identifies the contact when the electrical connection occurs between the tip and the surface of the workpiece. Alternatively, it is advisable to automatically measure the strain variation during the detection of the zero depth surface; when the end mill slowly touches the material surface, the strain gages detect the strain variation and the system immediately stops the drilling operation.

3. Advancements in residual stress calculation

After acquisition of the relaxed strain values, the residual stresses need to be evaluated. First of all, it is necessary to choose between blind-hole and through-hole calculation and between uniform or non-uniform stress distribution along the depth. When the residual stress value is uniform along the depth, the ASTM E837-13a standard specifies that the formulas described in Section 8 of the standard be used. The standard suggests assuming the stress to be uniform only when prior information about the expected stress field is available or if a representative size of the magnitude of the residual stress is required.

Typical applications generally exhibit a non-uniform state of residual stresses. In this case, it is necessary to follow the instructions reported in Section 9 of ASTM E837-13a for the calculation of residual stresses along the depth. The standard provides the calibration matrices, derived by the integral method, that must be multiplied with the acquired strains to derive the stress values. The matrix coefficients, corresponding to each calculation depth, are dimensionless and almost independent of the material [1].

For the sake of simplicity, only the equation for the calculation of the combination stress P is shown below.

$$\left(\bar{\boldsymbol{a}}^T\bar{\boldsymbol{a}} + \alpha_P \bar{\boldsymbol{c}}^T\bar{\boldsymbol{c}}\right) \cdot \boldsymbol{P} = \frac{E}{1+\nu}\bar{\boldsymbol{a}}^T\boldsymbol{p} \tag{1}$$

It is important to point out that dependence of the stresses on the Poisson's ratio, as shown in Eq. 1, is simplified [13].

All the coefficients reported in the calibration matrices A and B, for a non-uniform stress field, are strictly related to the nominal hole diameter (D_N) of 2 mm. If the diameter of the drilled hole (D_0) differs from the nominal value, each matrix coefficient reported in the standard needs to be corrected using Eq. 2 reported below:

$$\bar{a}_{j,kNEW} = \left(\frac{D_0}{D_N}\right)^2 \bar{a}_{j,k} \tag{2}$$

The dependency of the coefficients of the calibration matrix on the drilled hole diameter, as expressed in Eq. 2, is approximated as explained by Alegre et al. [7]. For example, using a rosette with a strain gage circle diameter (D) of 5.13 mm, the nominal hole diameter (D_N) is equal to 2 mm and the allowed diameter of the drilled hole (D_0) can vary from 1.88 to 2.12 mm. The ratio $(D_0/D_N)^2$ ranges between 0.88 and 1.12.

Recently, some developments have been carried out to overcome the limits of the ASTM E837 standard previously described and to take into account other parameters affecting the results that are not considered in the standard.

Beghini et al. [8, 9] of the University of Pisa introduced a generalized integral method based on the analytical definition of influence functions. The method is substantially an evolution of the Integral Method and it also overcomes the limitation of the ASTM E837 standard regarding the maximum allowable value of

Parameters	Limitations	
	ASTM E837 for non-uniform stress field calculation [1]	**Generalized integral method, based on the influence functions [8, 9]**
Hole diameter	Approximated correction if hole diameter differs from the nominal value used for the evaluation of the calibration coefficients	No limitation
Poisson's ratio	Approximated correction if Poisson's ratio differs from the value used for the evaluation of the calculation coefficients	Not applicable if outside the range 0.25–0.45
Rosette geometry	Fixed only for the strain gage rosette (A, B and C) reported in the standard.	Variable. Strain gage circle diameter (D), grid length (G_L) and width (G_W) can be used as input parameters
Hole eccentricity radius	Eccentricity correction is not included. Is considered acceptable if lower than 0.004D	No limitation
Workpiece thickness	<0.2D thick (uniform stress) >1.0D thin (uniform/non-uniform stress)	No limitation
Hole-bottom chamfer	Not considered	Considered (for a specific end mill geometry)

Table 1.
Comparison between the limitations of the ASTM E837-13a method and those of the generalized integral method based on the influence functions.

eccentricity. Using the Influence functions approach, it is also possible to include the real dependency of the Poisson's ratio and the diameter of the drilled hole on the calculated stress. In detail, the proposed methodology is based on analytical influence functions relating the measured relieved strains to the residual stress by means of integral equations. By processing the results of accurate finite element simulations, continuous analytical influence functions are produced.

The generalized integral method is more universal compared to the ASTM E837 standard and is currently the most suitable to include the influence parameters not considered in the standard and therefore to overcome its limitations. With this calculation method, it is possible to take into account other influence parameters, such as hole bottom chamfer and intermediate thickness.

Table 1 provides a comparison of the limitations of the ASTM E837 standard and of the generalized integral method based on the influence functions.

4. Generalized integral method

The ASTM E837 standard (Section 9.3) uses the integral method, including the Tikhonov regularization, to calculate non-uniform residual stresses. The residual stresses for each hole depth j are computed by solving the following matrix equations:

$$\left(\overline{\boldsymbol{a}}^T\overline{\boldsymbol{a}} + \alpha_P \boldsymbol{c}^T\boldsymbol{c}\right)\cdot \boldsymbol{P} = \frac{E}{1+\nu}\boldsymbol{p} \tag{3}$$

$$\left(\overline{\boldsymbol{b}}^T\overline{\boldsymbol{b}} + \alpha_Q \boldsymbol{c}^T\boldsymbol{c}\right)\cdot \boldsymbol{Q} = E\boldsymbol{q} \tag{4}$$

$$\left(\overline{\boldsymbol{b}}^T\overline{\boldsymbol{b}} + \alpha_T \boldsymbol{c}^T\boldsymbol{c}\right)\cdot \boldsymbol{T} = E\boldsymbol{t} \tag{5}$$

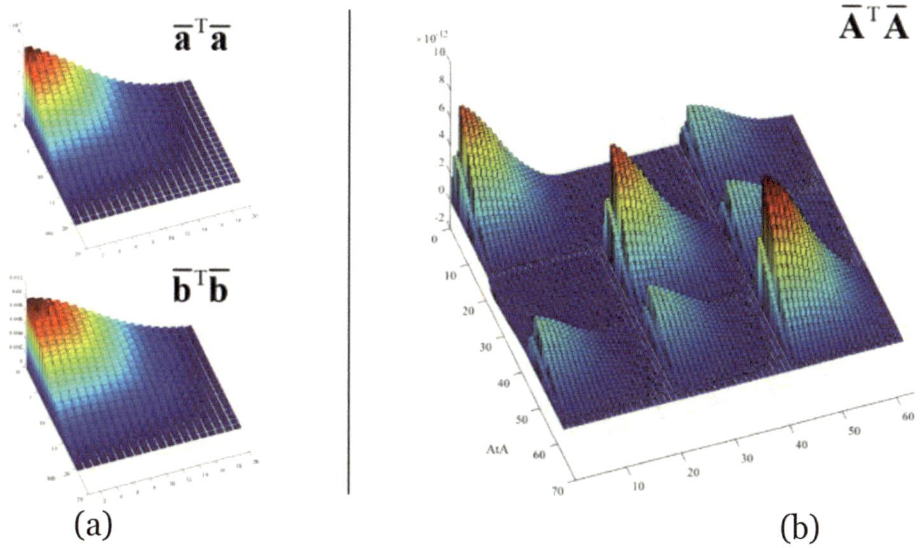

Figure 3.
(a) Two calibration matrices $\bar{a}$ and $\bar{b}$ for the ASTM E837 test method. (b) Calibration matrix A, composed of 9 blocks, for the generalized integral method.

Eqs. (3)–(5) are applicable only in the case of concentric holes, where it is possible to decouple the strain components.

Beghini et al. [8, 9] and, more recently, Barsanti et al. [13] extended the Integral Method by including a correction for the eccentricity of the hole with respect to the strain-gage rosette.

For this general problem, no symmetry can be used and no advantage is obtained by separating stress and relieved strain in equibiaxial and shear components. Therefore, the problem will be solved using the Cartesian reference system of the rosette.

The relationship between the strain and the stress can be re-written as reported below:

$$\left(\bar{A}^T\bar{A} + \alpha C^T C\right) \cdot S = E\bar{A}^T e \tag{6}$$

where $\mathbf{S} = \left(\sigma_x^{(1)}, \sigma_y^{(1)}, \tau_{xy}^{(1)}, \dots, \sigma_x^{(k)}, \sigma_y^{(k)}, \tau_{xy}^{(k)}\right)^T$ is the vector of the stress components; $e = \left(\varepsilon_1^{(1)}, \varepsilon_2^{(1)}, \varepsilon_3^{(1)}, \dots, \varepsilon_1^{(k)}, \varepsilon_2^{(k)}, \varepsilon_3^{(k)}\right)^T$ is the vector of strain reading; $\bar{A}$ = generalized matrix of calibration coefficients.

S and *e* have different sizes compared to the standard ASTM E837 approach. In this different formulation, the two vectors are defined using a $3\,k \times 1$ arrangement with a block structure of 3-elements.

The matrix $\bar{A}$ has a size of 3 k × 3 k and implicitly includes the dependency of the calculated stress on the hole diameter, type and dimension of the strain gage rosette (including the gage circle diameter, length and width of the strain grids), Poisson's ratio, eccentricity and, if applicable, hole bottom chamfer and thickness of the workpiece.

The equations reported above include also the Tikhonov regularization, as in the ASTM E837 standard.

As for the strain and the stress components, also the new matrix of calibration coefficients is defined for blocks of 3 × 3 elements as reported below:

$$\overline{A} = \begin{bmatrix} A_{11}^{(11)} & A_{12}^{(11)} & A_{13}^{(11)} & & & 0 & 0 & 0 \\ A_{21}^{(11)} & A_{22}^{(11)} & A_{23}^{(11)} & \cdots & \cdots & 0 & 0 & 0 \\ A_{31}^{(11)} & A_{32}^{(11)} & A_{33}^{(11)} & & & 0 & 0 & 0 \\ & \vdots & \ddots & . & & & \vdots & \\ & \vdots & . & \ddots & & & \vdots & \\ A_{11}^{(k1)} & A_{12}^{(k1)} & A_{13}^{(k1)} & & & A_{11}^{(kk)} & A_{12}^{(kk)} & A_{13}^{(kk)} \\ A_{21}^{(k1)} & A_{22}^{(k1)} & A_{23}^{(k1)} & \cdots & \cdots & A_{21}^{(kk)} & A_{22}^{(kk)} & A_{23}^{(kk)} \\ A_{31}^{(k1)} & A_{32}^{(k1)} & A_{33}^{(k1)} & & & A_{31}^{(kk)} & A_{32}^{(kk)} & A_{33}^{(kk)} \end{bmatrix} \tag{7}$$

Figure 3 compares the visual interpretation of the calibration matrices of the ASTM standard and that of the new generalized calibration matrix.

5. The main sources of errors and limits of applicability in the hole drilling method

The hole-drilling method has some typical sources of error that can influence the accuracy of the measurements. The ASTM E837 standard identifies the maximum values of these errors for the validity of the test (limits of applicability) without providing recommendations on how to correct them, as reported in **Table 2**.

Some of these errors, for example, eccentricity and hole bottom chamfer, can be generated by external sources such as the operator, testing condition or the drilling process. In other cases, the limits of applicability are directly connected with the test method, as in the case of intermediate thickness of the specimen or the local plasticity effect.

Table 2 also shows the bibliographic sources dealing with possible methods of error correction. The cases of uniform and non-uniform stress distribution are analyzed. In the case of uniform stress distribution, both the thin workpiece and the thick workpiece are considered.

The following sections examine the errors mentioned above in more detail and consider possible corrections.

Typical source of error	ASTM E837 limits of applicability	Suggested correction		
		Uniform		Non-uniform
		Thin	Thick	/
Hole eccentricity	Eccentricity radius within 0.004D	[12] [10, 11]	[8, 9] [10, 11]	[8, 9, 13, 14] [10, 11]
Intermediate thickness(s)	Thickness $s \leq 0.2D$ or $s \geq 1.0D$	[8, 9, 15, 16]		[8, 9, 16]
Hole-bottom chamfer	Not provided	Not applicable	[8, 9, 17]	[8, 9]
Local plasticity	Magnitude of the stresses $\leq$50% of σ_Y—Thin specimen $\leq$80% of σ_Y—Thick specimen	/	[20]	/

Table 2.
Typical sources of errors, ASTM E837 limit of applicability and current state-of-the-art of correction methodologies.

6. Eccentricity error: description and possible corrections

The eccentricity between the drilled hole and the strain gage circle greatly influences the strain measurements. The ASTM standard requires a near perfect concentricity between the drilled hole and the rosette and prescribes an allowable eccentricity value that depends on the dimension of the strain gage rosette (0.004D). Using a standard rosette with a gage circle diameter D = 5.13 mm, the maximum allowable eccentricity is 0.02 mm. This limit increases (0.04 mm) or decreases (0.01 mm) using bigger (approx. D = 10.26 mm) or smaller types of rosette (approx. D = 2.56 mm).

As shown in **Figure 4**, the eccentricity error is influenced by the eccentricity coordinates e_x and e_y and by the parameters of the strain gage rosette (D, G_L and G_W). The type B rosette generally shows a higher sensitivity to eccentricity errors compared to type A rosettes; this can be explained by the orientation of the gage grids, which are concentrated only in the first quadrant (0°, 45°, 90°), instead of in the first and third quadrant (0°, 90°, 225°).

For these reasons, the correction of eccentricity errors requires accurate determination of the position of the drilled hole in the reference system of the strain gage rosette; eccentricity can be measured by a special procedure using the drilling system microscope. Using a digital microscope, the eccentricity coordinates can be easily obtained by image analysis techniques.

As shown in **Figure 5**, the sensitivity of the grids is directly influenced by the hole eccentricity. When the eccentricity has the same direction as the grid, if the hole is closer to the grid the absolute value of the relaxed strain is greater. On the contrary, if the eccentricity has a transverse direction with respect to the grid, then a portion of the grid has a greater sensitivity, while the other portion has a lower sensitivity; this implies, by symmetry, that the error is almost compensated [13].

The eccentricity correction can be done using strain gage rosettes with special configurations or using correction algorithms.

The studies of Beghini et al. [10] and of Nau et al. [11] introduced the correction of eccentricity using a special six-grid rosette and an eight-grid rosette respectively. Both the rosettes are produced by HBM and make it possible to compensate the first-order of eccentricity error. However, the corrections based on special strain gage geometries do not correct the higher order eccentricity errors (higher than

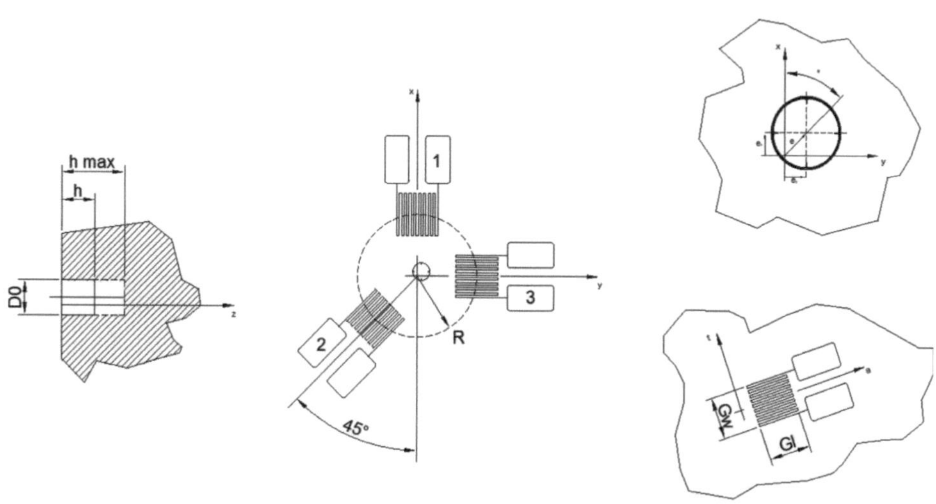

Figure 4.
Eccentricity in hole-drilling measurements.

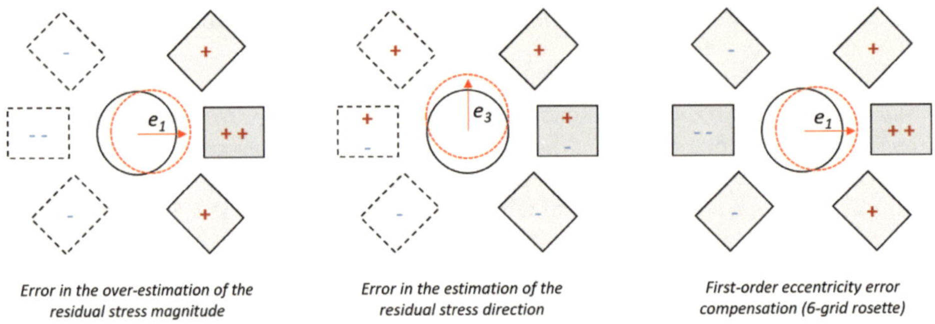

Figure 5.
Grid sensitivity to longitudinal and transversal eccentricity and first-order eccentricity error compensation using a 6-grid rosette.

0.2 mm), for which rosettes are required with bigger dimensions, a higher number of grids and higher costs.

Regarding the correction algorithms, the first solution was provided analytically by Ajovalasit et al. [12] for uniform stress in thin workpieces.

Beghini et al. [8, 9] provided a complete solution for blind holes using a generalized integral method based on the influence functions for non-uniform calculations (Section 4). According to this approach, the strain field is computed starting from a database of numerical solutions in which the eccentricity is simply introduced as a geometry parameter; this has the advantage of taking into account the whole effect of eccentricity. Recently, Barsanti et al. [13] proposed a simplified approach for the analytical correction of the first-order eccentricity errors in calculated stresses.

Peral et al. [14] has also proposed a correction approach applied to acquire strains.

7. Intermediate thickness limitation: description and possible correction

The ASTM E837 method defines the application ranges concerning the thickness of the workpiece under testing. The measurements can be carried out on "thin" or "thick" workpieces, the thickness of which depends on the size of rosette. For a "thin" workpiece, the thickness should be less than 0.20D (for type A and type B rosettes) and the stresses are evaluated according to the "uniform stress calculation". For a "thick" workpiece the thickness should be greater than 1.0D (for type A and type B rosettes) and the standard provides the calculation methods for uniform and non-uniform stress distributions.

The range of thicknesses between 0.2D and 1.0D, defined as intermediate thickness, is outside the scope of the ASTM standard. Using a strain gage rosette with a gage circle diameter D = 5.13 mm, the intermediate thickness is identified in the range between 1 and 5.13 mm. Clearly, on varying the diameter of the strain gage circle, also the range of the intermediate thickness varies.

Unfortunately, intermediate thickness is common in several types of engineering applications, as in aerospace, motor sports and energy production.

This limitation in the ASTM standard can be explained by analysis of the behavior of stress response if a hole is made in an intermediate thickness specimen.

In the case of thick workpieces (s > 1.0D) the influence coefficients are independent of the thickness and they can be obtained by an FE model in which the hole is produced in a virtually semi-infinite body. In the case of thin workpieces

($s \leq 0.2D$) the plane stress solution holds, the in-depth residual stress gradient is neglected and the through-hole method is applied; the influence coefficients for the thin plates can be directly deduced by Kirsch's solution of a membrane with a circular hole.

For the intermediate thickness case ($0.2D < s \leq 1.0D$), out-of-plane bending occurs (**Figure 6**) and this affects the calibration coefficients ***a*** and ***b*** defined in the ASTM standard. The calibration coefficients will depend directly on the workpiece thickness s, which becomes the new parameter.

A preliminary solution to this effect for the case of a uniform stress field was proposed by Abraham and Schajer [15]. They provide an analytical model of the calibration coefficients $\bar{a}, \bar{b}$ for intermediate thickness, as a function of workpiece thickness, hole diameter and hole depth.

Recently, Beghini et al. [16] described a procedure for the evaluation of non-uniform residual stress for the intermediate thickness range. The authors define two equations (one for the coefficients $\bar{a}_{j,k}$ and one for the coefficients$\bar{b}_{j,k}$), that adequately reproduce the thickness dependency of all the ASTM E837 calibration coefficients, for calculation of non-uniform stresses (**Figure 7**).

As reported in **Figure 7**, the dependency of the thickness is greater in the first part of the intermediate area (from 0.1D to 0.5D) and is less if the thickness is higher.

Moreover, a recent development of the generalized integral method based on the Influence Functions [8, 9] has introduced a new database of numerical solutions that takes thickness into account as an input parameter. The numerical database is

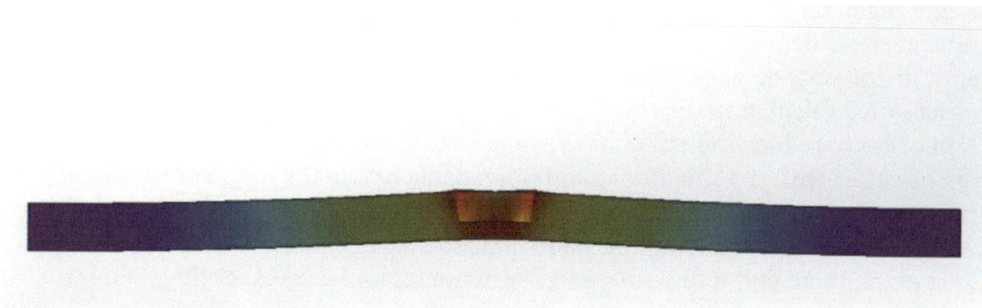

Figure 6.
Localized bending caused by hole-drilling in an "intermediate" thickness specimen.

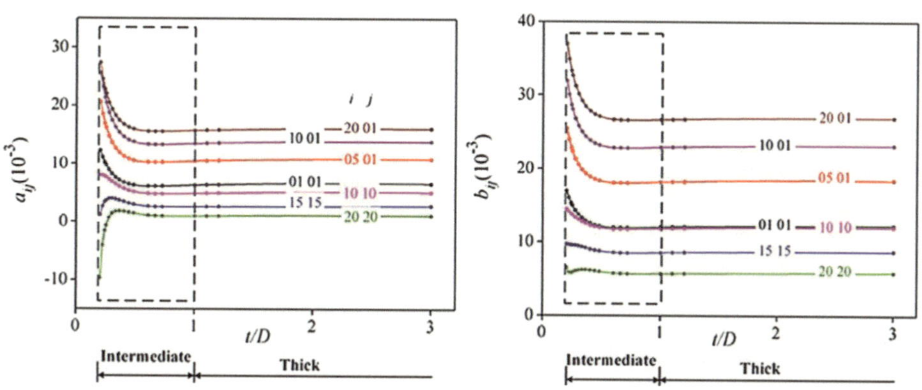

Figure 7.
Calibration coefficients as a function of plate thickness: for a_{ij} (left) and for b_{ij}, (right) coefficients in the matrices (from Beghini et al. [16]).

based on 5 different thicknesses (2.7 D, 1.0 D, 0.6 D, 0.3 D, 0.2 D); once the thickness is defined, the displacements are interpolated between the two closest available thickness values.

8. Hole bottom-chamfer error: description and possible correction

The hole-drilling method is based on the theoretical assumption that the drilled hole is perfectly cylindrical at any drilling increment. Perfect cylindrical holes are used in finite element models, by various authors, for the determination of the calibration coefficients.

The ASTM standard makes some recommendations regarding the geometry of the end mill in relation to both the radial clearance angles of the cutting edges on the end face of the cutting tool (<1°) and the taper angle (<5°). These requirements were introduced by the standard in order to avoid any ambiguity in determination of the depth and measurement of the hole diameter.

Unfortunately, the carbide inverted-cone end mills used for performing the hole-drilling measurements could have a small chamfer at their cutting extremities that generates a small chamfer in the bottom of the hole (**Figure 8**). This small chamfered extremity of the end mill reduces wear and facilitates chip ejection during drilling.

This chamfer influences the strain signals and consequently the calculation of residual stresses. The effect of the hole bottom chamfer has a higher impact on the first depth increments where the chamfer of the end mill generates a hole with a smaller diameter than the nominal diameter. Furthermore, in the case of non-uniform stress distribution, the geometric variation in the hole shape in the first depth increments determines errors not only in the first calculation depths but also in successive calculation depths.

In order to reduce the effect of the hole-bottom chamfer, it is advisable to use a type of cutter with the smallest chamfer available or use the high-speed orbital drilling technique. The method is based on the orbital movement of the end mill as it advances. While producing the same diameter as a center-drilled hole, this technique employs an end mill with a smaller diameter and consequently creates a smaller bottom chamfer.

If the above is not possible, it is necessary to correct the errors generated by the presence of the chamfer.

A first solution for this correction was proposed by Scafidi et al. [17] carrying out an analysis based on the Boundary Element Method. By introducing the gage circle diameter, the drilled hole diameter and the bottom-hole fillet radius, the authors developed a method based on the correction of acquired strains. Subsequently,

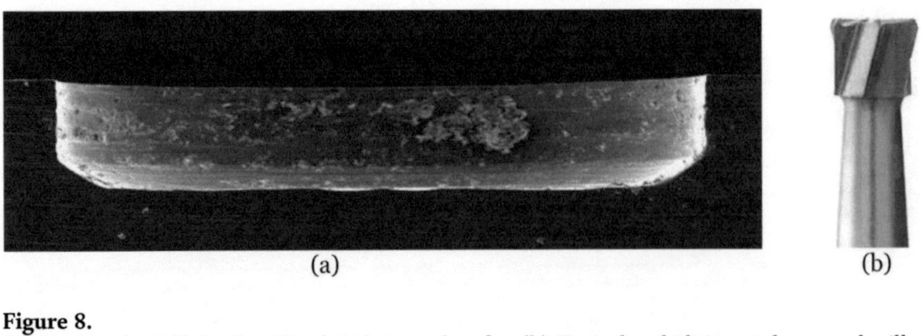

(a) (b)

Figure 8.
(a) Section of a drilled hole with a hole-bottom chamfer, (b) Typical carbide inverted-cone end mill used for the hole drilling method.

Blödorn et al. [19] recalculated the ASTM E837 coefficient for blind uniform stress using an FEM model with a hole bottom chamfer.

More recently, the generalized integral method based on the influence functions [8, 9] has been enriched with a new database of displacements, which considers the chamfer as a new geometrical parameter of the finite element model.

For a certain value of the ratio between the height of the hole chamfer and the radius of the drilled hole, this methodology allows the correction of calculated stress for blind holes and non-uniform stress distributions.

Figure 9 shows the finite element model in which the hole bottom chamfer was simulated to evaluate its influence.

The presence of the hole-bottom chamfer influences the calculation of the stresses.

Figure 10 gives an example of the influence of a hole bottom chamfer on the reconstruction of a pure shear stress distribution. In the first part of the depth of the analysis, it is clearly seen that the chamfer determines an under-estimation of the actual stress, especially in the first depth increments. On the contrary, in the second part of the depth of the analysis, the results show an over-estimation of the calculated stresses.

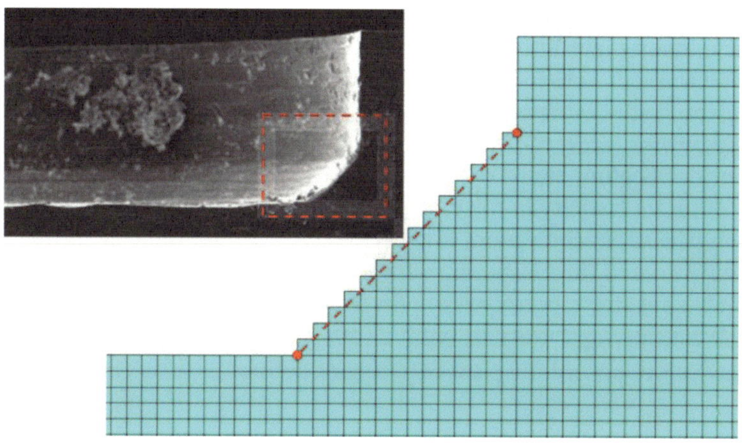

Figure 9.
Finite Element Model used for the evaluation of the calibration coefficients considering the presence of the hole-bottom chamfer.

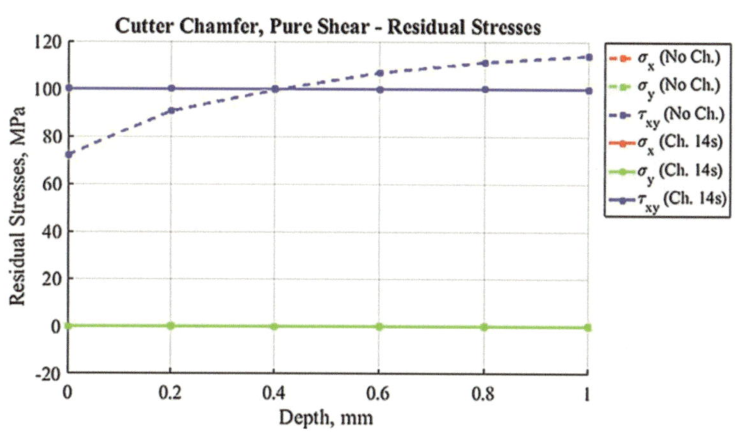

Figure 10.
Residual stresses in the case of pure shear stress, with (dashed line) and without (solid line) the hole-bottom chamfer.

9. Local plasticity limitation: description and possible correction

The ASTM E837 standard reports that satisfactory results can be achieved when measured residual stresses do not exceed about 80% of the yield stress in the case of "thick" workpieces and 50% in the case of "thin" workpieces.

The need for these stress limits is explained by the stress concentration generated by the drilled hole. When a hole is drilled on a loaded workpiece, it generates a stress concentration in the area around the hole. The magnitude of the stress concentration depends on several parameters including the diameter of the drilled hole, the load orientation and the distance of the strain gage grids from the hole. If the stress level is high, localized plastic deformation occurs around the hole, which generates larger overall surface strains (**Figure 11**).

The hole-drilling method requires that the strain gage grids be placed really close to the hole. For this reason, if local plasticity occurs, it may be that the strain measured by the gage is the arithmetical sum of the linear elastic strains and the plastic strains.

In any case, "thick" workpieces are less sensitive to the plasticity effect. This is due to the presence of material in the lower part of the blind hole determining a local reinforcement and reducing the stress concentration factor [3]. This explains the higher measurement limit in the case of a blind hole (80% of σ_Y) compared to the case of a through hole (50% of σ_Y).

Few research studies have been published on this topic to provide possible corrections for this error.

The work of Beghini et al. [20] provides a numerical procedure for correcting the effect of local plasticity in the case of a blind hole for uniform stress calculation. To carry out the stress correction, it is necessary that both the yield stress and the stress–strain curve in the plastic region are defined.

The equivalent stress, corrected to take into account the presence of plasticity σ_{eq}, is evaluated considering the elastic equivalent stress $\sigma_{eq,i}$, the yield stress σ_Y of the material under testing and the plasticity factor f defined as following:

$$f = \frac{\sigma_{eq} - \sigma_{eq,i}}{\sigma_Y - \sigma_{eq,i}} \tag{8}$$

The correction algorithm obviously considers the geometry of the strain gage rosette and therefore the authors provide the calculation coefficients for several strain gage rosettes available on the market.

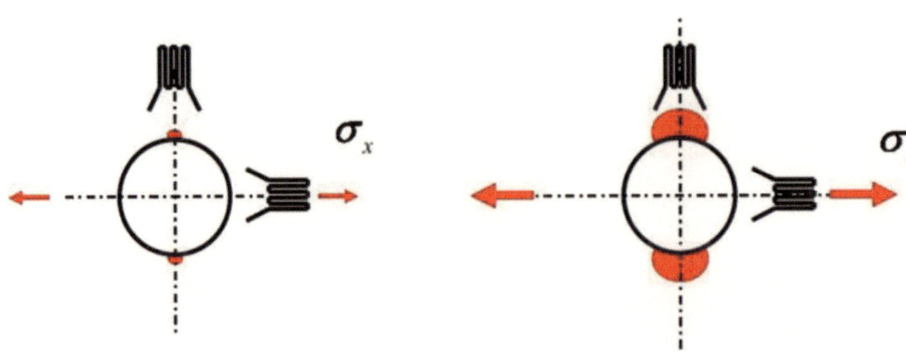

Figure 11.
Local plasticity areas with low applied loads (left side) or high loads (right side).

The previous parameters and FE results are used for the evaluation of the elastically evaluated plasticity factor f_{el}, which is expressed through bivariate polynomials, as a function of the parameters W and μ

$$f_{el} = f + Wf^{\mu} \tag{9}$$

Nobre et al. [21] provide a similar approach for the estimation of the plasticity factor f. The material characteristics are taken into account by measuring the variation of Vickers hardness, which estimates the material strain hardening due to the increase of plastic deformation.

Plasticity generates a non-linearity on strain measurements.

Beghini et al. [22] propose a special 4-grid strain gage rosette for the correction of the plasticity effect, which is available on the market (HBM). The correction is valid for the standard 3-grid rosettes only if the perpendicular grids are oriented in the directions of the principal strains.

10. Evaluation of uncertainty

The evaluation of uncertainties associated with measurement of residual stresses by the hole-drilling method is a topic that has been little investigated. The evaluation, mainly in the case of non-uniform stress fields, involves a large number of parameters from different sources, including the properties of the materials under testing, the strain readings and the hole execution methods.

Standard ASTM E837-13a [1] contains only some basic information about precision and bias associated with the hole-drilling measurement method, mainly in the case of uniform stress calculation. In fact, the standard states that the bias associated with a residual stress measurement by the hole-drilling method is less than ±10% when dealing with uniform residual stresses. Based on the results of round-robin test programs, the precision (random error) is such as to give a standard deviation of ±14 MPa for AISI 1018 carbon steels and a standard deviation of ±12 MPa for type AISI 304 stainless steels. The standard also reports that the uncertainties in the case of non-uniform stress measurements are expected to be much larger than for uniform stress measurements.

One of the first papers on the subject of evaluation of uncertainty was published by Oettel [25] (UNCERT Code of Practice 15). The work proposes an approach for the evaluation of hole-drilling uncertainty in the case of uniform stress fields and takes into account typical errors in the determination of material properties, errors in the measurement of acquired strains, the hole diameter and the influence of calculation coefficients. The code of practice can be applied only to uniform residual stress calculation equations based on ASTM E837-95 and cannot be used with the current version of the ASTM standard [1].

Scafidi et al. [26] further developed this methodology by applying it to the recent version of the average uniform stress calculation and considering additional parameters, such as the step-by-step drilling depth.

Regarding evaluation of uncertainty in the case of non-uniform stresses, the first approach was provided by Schajer et al. [24] based on the Integral method. They consider a number of input estimates including the properties of materials (i.e. Young's modulus), strain readings, hole diameter and hole depths. The uncertainty components have statistical normal distributions with zero mean and are independent of each other and each one is linearly combined.

The uncertainties of the measured strains are considered as an input, but fixed for each step.

More recently a new approach was proposed by Peral et al. [27], based on a Monte Carlo analysis of the influence of the main parameters affecting the measurements. The methodology takes into account a higher number of parameters compared to the approach proposed by Schajer. In particular, the uncertainty components due to Poisson's ratio and identification of the zero-depth are also considered. The authors demonstrated that their method is comparable with the approach of Schajer et al. [24], showing generally more conservative results although in good agreement.

SINT Technology recently developed another approach to evaluation of uncertainty, based on the GUM methodology [23], and it is implemented in the calculation software EVAL 7.

Before calculation of uncertainties, all possible systematic errors are corrected, in particular those determined by eccentricity (Section 6), intermediate thickness (Section 7), hole bottom chamfer (Section 8) or local plasticity in the case of uniform stress calculation (Section 9).

The uncertainties determined by the following input parameters are considered: Young's modulus, Poisson's ratio, hole diameter, accuracy of the strain measurement system, zero depth offset error, depth of drilling increments.

This approach can be applied to all available calculation methods including the ASTM standard: clearly, the generalized integral method approach (Section 4) is preferable as it allows several systematic errors to be corrected (Sections 6–8).

The functional relationship *f* that connects the output estimate to the inputs and the measurements can be written, for each calculation step *j*, as:

$$\sigma_{MIN,MAX}, \beta = f\left(E, \nu, D_0, \varepsilon_{1,j}, \varepsilon_{2,j}, \varepsilon_{3,j}, z_j, z_0\right) \tag{10}$$

where E is the Young's modulus of the material; ν the Poisson's ratio; D_0 the diameter of the drilled hole; $\varepsilon_{1,j}, \varepsilon_{2,j}, \varepsilon_{3,j}$ the Readings of the strain gage grids for the step *j*; z_j the depth advance for the step *j* and z_0 the depth error during the zero-depth determination.

The reading of the strain gage grids, for each calculation step *j* and for each channel *y*, is derived from the following parameters:

$$\varepsilon_{1,j}, \varepsilon_{2,j}, \varepsilon_{3,j} = f(K_x, (\Delta V/V)_{j,x} \tag{11}$$

where K_x is the Gage Factors of the strain gage for each grid *x*; $(\Delta V/V)_{j,x}$ is the electrical signal output for each channel *x* and for each step *j*.

The uncertainty on the electrical output $(\Delta V/V)_{j,x}$ depends on the strain gage amplifier technical specifications, such as class of accuracy or resolution, linearity and noise to signal ratio. This information can be obtained from the technical datasheet of the strain gage amplifier.

Also the uncertainty on the measurement of the drilled hole D_0 and the depth increments z_j, for each *j* step, are evaluated starting from the technical specification respectively of the dial gages used for the hole measurements and the mechanical drilling unit that makes the hole.

The uncertainty component due to the temperature variation during testing is considered negligible as the strain gages are self-compensated and a three-wire half-bridge connection is adopted to minimize the effect of temperature on the cables. Also the uncertainties due to the Influence Functions are considered negligible. **Table 3** shows the typical input parameters taken into account for the evaluation of the uncertainty, along with the type of statistical distribution.

Assuming that all the input quantities are independent, the combined standard uncertainty, for each calculation step *j*, is given by:

Input estimates x_i	Description	Sub-input estimates	Distribution	Origin
E	Young's Modulus of the materials under testing	/	Rectangular	Material datasheet
ν	Poisson's ratio of the materials under testing	/	Rectangular	Material datasheet
D_0	Diameter of the drilled hole	/	Normal	Resolution of the dial gages
				Max. error of the dial gages
				Repeatability of the dial gages
$\varepsilon_{1,j}, \varepsilon_{2,j}, \varepsilon_{3,j}$	Gage factor of the strain gage grids	K_x for each channel x	Normal	Uncertainty declared on the strain gage datasheet
	Electrical output of each strain gage grid	$(\Delta V/V)_{j,x}$ for each channel x for each step j	Normal	Class of accuracy of the strain gage amplifier
				Linearity of the strain gage amplifier
				Resolution of the strain gage amplifier
				Noise of the strain gage amplifier
z_j	Depth increment	/	Normal	Resolution of the hole-drilling mechanical device
				Max. error between two consecutive steps of the hole-drilling mechanical device
z_0	Zero-depth error	/	Rectangular	Datasheet of the hole-drilling mechanical device

Table 3.
Parameters used for the uncertainty evaluation and distribution of probability.

$$u_c(y) = \sqrt{\sum_{I=1}^{N} \left(\frac{\partial f}{\partial x_I}\right)^2 u^2(x_I)} \tag{12}$$

where y is the measurement result (output estimate); $u_C(y)$ is the combined standard uncertainty for measurement result; x_I the input quantity measurement (input estimate); $u(x_I)$ is the standard uncertainty for each input quantity; $I = 1, \dots, N$ is the number of input quantities.

Finally, the expanded uncertainty U is obtained by multiplying the combined standard uncertainty $u_C(y)$ by a coverage factor k:

$$U = ku_C(y) \tag{13}$$

The result of a measurement is then conveniently expressed as:

$$Y = y \pm ku_C(y) \tag{14}$$

The advantage of this method is the capability to numerically evaluate, for each parameter and for each calculation depth, the first derivatives of the functional relation f (sensitivity coefficients) which, in the formula (12), multiply the standard uncertainty square $u(x_i)$ of each input estimate x_i.

This calculation procedure, which is implemented in the EVAL 7 software, requires the execution of a high number of stress calculations for the uncertainty evaluation related to hole drilling measurements.

In particular, considering a measurement carried out according to the ASTM standard using 20 calculation depths, the uncertainty evaluation requires the repetition and therefore the combination of the results obtained with 206 different stress calculations.

11. Experimental validation of the method on a four-point bending rig

The entire measurement chain and the testing parameters (rotational speed, type of end mill, feed rate and delay time) were verified using a special apparatus, developed by SINT Technology, which applies a known bending stress on a specially designed specimen.

The specimen is a flat rectangular cross section cantilever beam, fixed at one end, and loaded at the other end by means of a pneumatic actuator (**Figure 12**).

The material used for the specimen is aluminum alloy AW7075 T651. The maximum applied bending stress was approximately 25 MPa.

An approximately 1.90 mm hole diameter was drilled with 130 incremental drilling depths of 10 μm up to a 1.30 mm total depth. The rotational speed was approx. 400,000 rpm and the feed rate 0.2 mm/min. To prevent any interaction between the tip and the specimen, the cutter was fully raised for each drilling step. The diameter was accurately measured after drilling, for each test, with the microscope installed on the system and two dial gages, also to determine the residual eccentricity.

The (uniaxial) stress due to bending was easily obtained from the beam theory, **Figure 12b**, Eq. 15:

$$\sigma_{Be} = 6\,\frac{Fb}{wh^2} \tag{15}$$

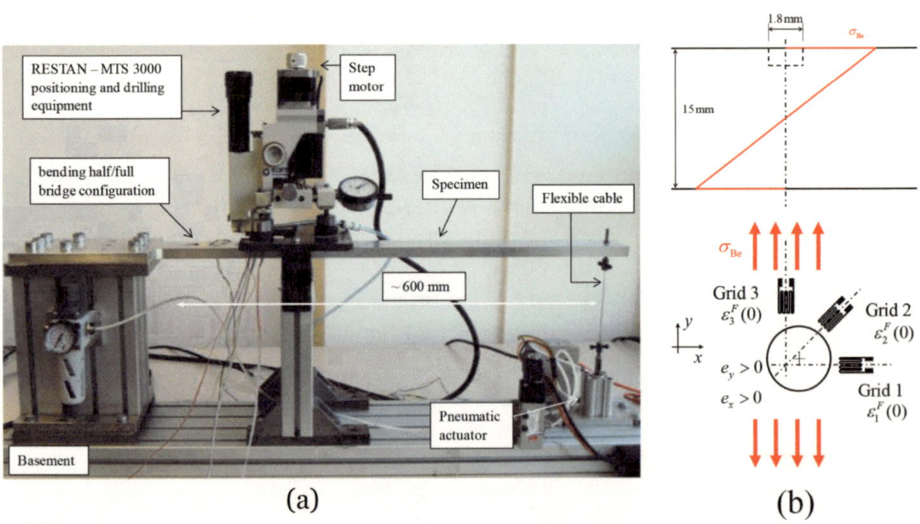

Figure 12.
(a) Bending test bench to simulate a known reference residual stress and (b) shows the linear distribution of the bending stress and orientation of the strain gage rosette grids and hole eccentricity definition.

where b is the distance between the load axis and the rosette strain gage center, wh^2 is the width and height of the beam cross section, and F is the load imposed by the pneumatic actuator.

A known load was used for determining the properties of the material. In fact, the elasticity parameters (Young's modulus E and Poisson's ratio ν) of the material were measured before drilling by applying a preliminary bending load before drilling.

Grid 1 of each strain gage should be aligned with the beam axis. The manual strain gage installation unavoidably introduces a misalignment. However, the angle between grid 1 and the beam axis can be found from Eq. 16 (accurate approximation for small values of γ):

$$\gamma = \frac{1}{2} \cdot \frac{\varepsilon_1^F(0) - 2\varepsilon_2^F(0) + \varepsilon_3^F(0)}{\varepsilon_1^F(0) - \varepsilon_3^F(0)} \tag{16}$$

The measured strains need to be decoupled in order to deduce the relaxed strain due to the bending stress. The relaxed strains due to the residual stresses and the relaxed strains due to the bending stresses are obtained as:

$$\begin{aligned} \varepsilon_i^{RS}(z_j) &= \varepsilon_i(z_j) \\ \varepsilon_i^{Be}(z_j) &= \varepsilon_i^F(z_j) - \varepsilon_i(z_j) - \varepsilon_1^F(0) \end{aligned} \tag{17}$$

Strain $\varepsilon_i^F(0)$ needs to be subtracted in the second member of Eq. 17 since the relaxed strains are defined as the effect of introduction of the drilled hole, therefore they need to be zero at zero depth. Finally, the experimental data are the bending relaxed strains as a function of hole depth increments $\varepsilon_i^{Be}(z_j)$.

12. Test results and analysis

The following testing conditions were adopted during the measurements.

- Surface bending stress (σ_{Be}): 24.8 MPa
- Strain gage rosette: CEA-062UM-120
- Measured eccentricity radius: 0.02 mm
- Measured eccentricity angle: 135°

The following parameters were then used for the stress calculation and for the uncertainty evaluation:

- Young's modulus (E): 71000 $\pm$ 3550 MPa
- Poisson's ratio (ν): 0.33 $\pm$ 0.01
- Hole diameter (D_0): 1.88 $\pm$ 0.01 mm
- Gage factor uncertainty (K): 1%
- Uncertainty on the maximum electrical output ($\Delta V/V$): $\pm$0.50 μm/m

- Zero-depth uncertainty (z_0): ± 0.005 mm
- Depth measurement uncertainty (z): ± 0.01 mm

After performing the drilling tests, the relaxed strains were imported into the EVAL 7 calculation software developed by SINT Technology. The calculation of non-uniform stresses was carried out according to the following two methods: the original ASTM 837-13a standard and the generalized integral method, based on the Influence Functions, by applying the algorithms described in the previous sections and correcting some systematic errors.

The extended features are shown in **Table 4**.

The stresses were calculated considering a distribution of 20 constant steps within 1 mm of depth.

Next, the bending stress distribution was calculated by the generalized integral method and then compared with the expected bending stress distribution.

Finally, based on the calculated stress curves, the uncertainty of measurement was evaluated considering the input quantities reported above (Section 12). The measurement uncertainties are expressed as standard uncertainties multiplied by a coverage factor equal to 2 (which in the case of normal distribution corresponds to a confidence level of about 95%).

Figure 13 compares the expected bending stresses with the stress components σ_x, σ_y and τ_{xy}, calculated from the interpolated relaxed strains with their associated uncertainty.

The purpose of the authors is to highlight the importance of the correction of each source of error, which is not contemplated in the ASTM E837 calculation. This has been achieved by showing the effects on the calculated stresses in the event that those corrections are not considered. For this reason, one by one, all the corrections have been deselected from the generalized integral method with all the active corrections.

Figure 14 shows the percentage error on σ_{Be} when the generalized integral method is not applied, due to the hole eccentricity, the combination of Poisson's ratio and the hole diameter, and the geometry of the strain gage rosette.

Regarding eccentricity, a maximum error of approximately 1.5% is committed, in the area closest to the surface. It is necessary to highlight that the eccentricity radius error, that affects this data, is similar to the maximum value tolerated by the ASTM 837-13a standard. In some real cases, due to the inexperience of the operator

Features	**Comparison**	
	ASTM E837-13a	**Generalized integral method, based on the influence functions**
Eccentricity correction	Not available	Available
Applicability to strain gage rosettes	Only to rosettes listed in the standard	To any rosette available in the market
Poisson's ratio correction	Approximate	Complete
Hole diameter correction	Approximate	Complete
Hole-bottom chamfer correction	Not available	Available
Intermediate thickness extension	Not available	Available
Tikhonov regularization	Available	Available

Table 4.
Comparison between the ASTM E837 & generalized integral method, based on the influence functions.

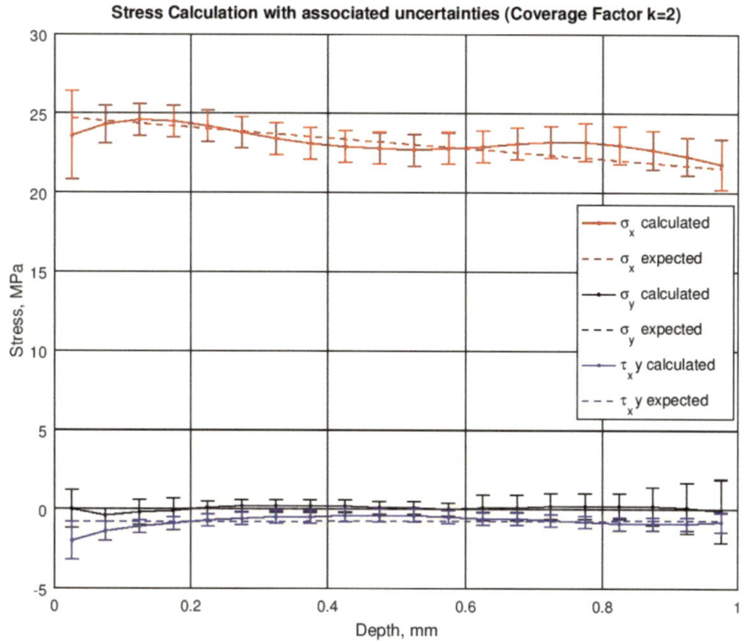

Figure 13.
Comparison between expected bending stresses and the calculated stress components.

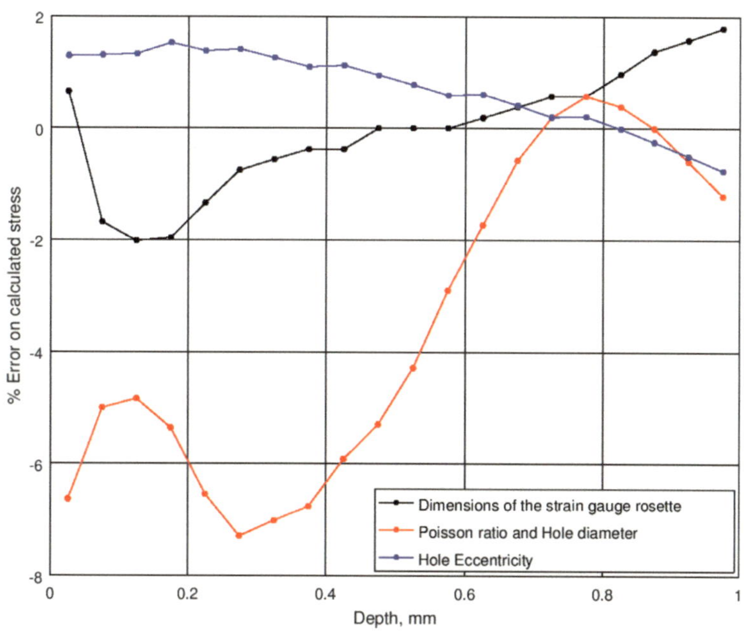

Figure 14.
Percentage error on σ_{Be} *when the generalized integral method is not applied.*

or to non-standard test conditions, the eccentricity radius can be higher than the limitation reported by the standard and, therefore, correction of eccentricity is essential for an accurate evaluation of residual stresses.

Regarding the influence of Poisson's ratio and the hole diameter, the maximum deviation is around 8%. Indeed, the calibration constants are not expressed as a function of Poisson's ratio and the diameter of the measured hole: only the approximate correction is provided. In this case, both the measured diameter (D = 1.88 mm) and the Poisson's ratio considered (n = 0.33 mm) are far from those used to generate the calibration matrices reported in the standard.

Finally, regarding the influence of the geometry of the strain gage rosette, the maximum deviation is approximately 2%. It represents the error due to use of a rosette that is different from the geometry envisaged in the standard. In this case, the rosette that was used is very similar to type B of the standard. In other cases, the errors may be higher.

13. Conclusions

The paper describes some improvements in the hole-drilling test method for the analysis of residual stresses, developed to increase accuracy. These improvements have been introduced to overcome some limitations and correct some errors that can derive from the direct application of the ASTM E837-13a standard.

To make calculation of the distribution of non-uniform stresses more accurate, the evaluation of residual stresses was carried out by applying the general method, based on the influence functions, proposed by Beghini et al. [8, 9]. This approach is more extensive with respect to the integral method proposed by Schajer [13, 14] and can include a dependency on a higher number of parameters. This more general approach avoids some errors and removes some limitations in the evaluation of non-uniform residual stresses deriving from the application of the ASTM E837-13a standard, which is based on the Integral method. Ultimately, the Integral method can be considered as a special case of the influence function method in which piecewise constant functions are used as the basis.

With this approach, some limitations of the standard can be overcome and, in particular, applicability of the hole-drilling test method is extended:

- For all strain gage rosettes available on the market, instead of just the A, B and C type rosettes [1],
- When eccentricity of the hole is greater than 0.004 D
- When the thickness s of the workpiece is between $0.2D < s \leq 1.0D$.

Furthermore, with this approach it will be possible to correct errors due to:

- hole eccentricity
- hole bottom chamfer
- approximate correction if the hole diameter and Poisson's ratio differ from the nominal value used for the evaluation of the calibration coefficients in ASTM
- local plasticity, only in the case of blind holes and uniform stress

The paper also describes developments in measurement instrumentation with the use of automatic systems instead of manual systems and a procedure for evaluating measurement uncertainties in the case of non-uniform distribution, based on the GUM method.

All the features reported above have been introduced in dedicated software for the evaluation of residual stresses and related uncertainty. Finally an experimental test, performed on a 4-point bending test rig, is described.

List of symbols

σ_x	residual stress normal component in the x direction [MPa]
σ_y	residual stress normal component in the y direction [MPa]
τ_{xy}	residual stress shear component in the xy plane [MPa]
σ_{MIN}	minimum principal residual stress [MPa]
σ_{MAX}	maximum principal residual stress [MPa]
β	principal angle [rad]
σ_Y	yield stress of the testing material [MPa]
D	diameter of the strain gage circle [mm]
D_0	diameter of the drilled hole [mm]
D_N	ASTM E837 nominal hole diameter [mm]
G_L	grid length [mm]
G_W	grid width [mm]
$\varepsilon_1, \varepsilon_2, \varepsilon_3$	strains acquired from the strain gage rosette [μm/m]
p, q, t	combination strain [μm/m]
P, Q, T	combination stress [MPa]
E	Young's modulus [MPa]
ν	Poisson's ratio
$\bar{a}, \bar{b}$	calibration constants used in the calculation of uniform stress
s	workpiece thickness [mm]
n	number of acquisition steps
j	number of hole depth steps
k	sequence number for hole depth steps
$\bar{\mathbf{a}}, \bar{\mathbf{b}}$	calibration matrix constants used in the calculation of non-uniform stress
c	Tikhonov regularization matrix
$\alpha_P, \alpha_Q, \alpha_T$	Tikhonov regularization factors
$\bar{a}_{j,k}, \bar{b}_{j,k}$	calibration matrix for isotropic and shear stresses
S	vector of the stress components
e	vector of the strain components
$\bar{A}$	generalized matrix of calibration coefficients
C	generalized Tikhonov regularization matrix
e_x, e_y	eccentricity component of the x and y directions
$\sigma_{eq,el}$	equivalent stress [MPa] corrected for plasticity effect
$\sigma_{eq,i}$	equivalent residual stress producing the onset of plasticity in the 2D case
f, f_{el}	plasticity factor calculated in plastic and elastic field
W, μ	coefficients for the plasticity correction
z_j	depth increment j
z_0	depth error during the zero-depth determination
K_x	gage factors of the strain gage for each grid x (x = 1,2,3)
$(\Delta V/V)_{j,x}$	electrical output reading for each grid x (x = 1,2,3) and for each depth increment j
y	measurement result (output estimate)
$u_C(y)$	combined standard uncertainty for measurement result

x_I	input quantity measurement (input estimate)
$u(x_I)$	standard uncertainty for each input quantity
I	number of input quantities

Author details

Emilio Valentini*, Lorenzo Bertelli, Alessio Benincasa and Simone Gulisano
SINT Technology S.r.l., Florence, Italy

*Address all correspondence to: emilio.valentini@sintechnology.com

References

[1] ASTM E837-13a. Standard Test Method for Determining Residual Stresses by the Hole-Drilling Strain-Gage Method. West Conshohocken, PA: ASTM International, West Conshohocken; 2013. Available from: www.astm.org

[2] Valentini E, Bertelli L, Benincasa A. Improvements in the hole-drilling test method for determining residual stresses in polymeric materials. Materials Performance and Characterization. 2018;7(4):446-464. DOI: 10.1520/MPC20170123 ISSN 2379-1365

[3] Schajer GS, Whitehead PS. Hole-Drilling Method for Measuring Residual Stress. Vermont, USA: Morgan & Claypool; 2018. 186 pp

[4] Schajer GS, Whitehead PS. Hole-drilling and ring coring. In: Schajer GS, editor. Chapter 2 in Practical Residual Stress Measurement Methods. Chichester, UK: Wiley; 2013. pp. 29-64

[5] Schajer GS. Measurement of non-uniform residual stresses using the hole-drilling method. Part I—Stress calculation procedures. Journal of Engineering Materials and Technology. 1988;**110**:338-343

[6] Schajer GS. Measurement of non-uniform residual stresses using the hole-drilling method. Part II—Practical application of the integral method. Journal of Engineering Materials and Technology. 1988;**110**:344-349

[7] Alegre JM, Díaz A, Cuesta II, Manso JM. Analysis of the influence of the thickness and the hole radius on the calibration coefficients in the hole-drilling method for the determination of non-uniform residual stresses. Experimental Mechanics. 2019;**59**:79. DOI: 10.1007/s11340-018-0433-0

[8] Beghini M, Bertini L, Mori LF. Evaluating non-uniform residual stress by the hole-drilling method with concentric and eccentric holes. Part I. Definition and validation of the influence functions. Strain. 2010;**46**: 324-336. DOI: 10.1111/j.1475-1305.2009.00683.x

[9] Beghini M, Bertini L, Mori LF. Evaluating non-uniform residual stress by the hole-drilling method with concentric and eccentric holes. Part II: Application of the influence functions to the inverse problem. Strain. 2010;**46**: 337-346. DOI: 10.1111/j.1475-1305.2009.00684.x

[10] Beghini M, Bertini L, Santus C, Benincasa A, Bertelli L, Valentini E. Validazione sperimentale di una rosetta a 6 griglie per ridurre l'errore di eccentricità nella misura delle tensioni residue. In: Congresso AIAS XXXIX; Maratea (PZ). 2010. ISBN: 8860930749

[11] Nau A, Scholtes B. Experimental and numerical strategies to consider hole eccentricity for residual stress measurement with the hole drilling method. Materials Testing. 2012;**54**(5): 296-303. DOI: 10.3139/120.110330

[12] Ajovalasit A. Measurement of residual stresses by the hole-drilling method: influence of hole eccentricity. Journal of Strain Analysis for Engineering Design. 1979;**14**:171-178

[13] Barsanti M, Beghini M, Bertini L, Monelli BD, Santus C. First-order correction to counter the effect of eccentricity on the hole-drilling integral method with strain-gage rosettes. The Journal of Strain Analysis for Engineering Design. 2016;**51**(6):431-443. DOI: 10.1177/0309324716649529

[14] Peral D, Correa C, Diaz M, et al. Measured strains correction for

eccentric holes in the determination of non-uniform residual stresses by the hole drilling strain gage method. Materials & Design. 2017;**132**:302-313. DOI: 10.1016/j.matdes.2017.06.051

[15] Abraham C, Schajer GS. Hole-drilling residual stress measurement in an intermediate thickness specimen. In: Experimental and Applied Mechanics. Vol. 4. Conference Proceedings of the Society for Experimental Mechanics Series. New York, NY: Springer; 2013. Available from: https://doi.org/10.1007/978-1-4614-4226-4_46

[16] Beghini M, Bertini L, Giri A, Santus C, Valentini E. Measuring residual stress in finite thickness plates using the hole-drilling method. The Journal of Strain Analysis for Engineering Design. 2019;**54**(1):65-75. DOI: 10.1177/0309324718821832

[17] Scafidi M, Valentini E, Zuccarello B. Effect of the hole-bottom fillet radius on the residual stress analysis by the hole drilling method. In: ICRS-8 The 8th International Conference on Residual Stress; Denver. 2008. pp. 263-270

[18] Simon N, Gibmeier J. Consideration of tool chamfer for realistic application of the incremental hole-drilling method. Materials Research Proceedings. 2017;**2**: 473-478. DOI: 10.21741/9781945291173-80

[19] Blödorn R, Bonomo L, Viotti M, Schroeter R, Albertazzi A Jr. Calibration coefficients determination through FEM simulations for the hole-drilling method considering the real hole geometry. Experimental Techniques. 2017;**41**:37. DOI: 10.1007/s40799-016-0152-3

[20] Beghini M, Bertini L, Santus C. A procedure for evaluating high residual stresses using the blind hole drilling method, including the effect of plasticity. The Journal of Strain Analysis for Engineering Design. 2010;**45**(4): 301-318. DOI: 10.1243/03093247JSA579

[21] Nobre JP, Kornmeier M, Scholtes B. Plasticity effects in the hole-drilling residual stress measurement in peened surfaces. Experimental Mechanics. 2018;**58**:369. DOI: 10.1007/s11340-017-0352-5

[22] Beghini M, Santus C, Valentini E, Benincasa A. Experimental verification of the hole drilling plasticity effect correction. Materials Science Forum. 2011;**681**:151-158. DOI: 10.4028/www.scientific.net/MSF.681.151

[23] JCGM 100. Evaluation of measurement data—Guide to the expression of uncertainty in measurement. ISO/IEC Guide 98-3. 2008

[24] Schajer GS, Altus EE. Stress calculation error analysis for incremental hole-drilling residual stress measurements. ASME Journal of Engineering Materials and Technology. 1996;**118**(1):120-126. DOI: 10.1115/1.2805924

[25] Oettel R. The determination of uncertainties in residual stress measurement (using the hole drilling technique). Code of Practice. EU Project No. SMT4-CT97-2165. Sept. 2000;**15**(1)

[26] Scafidi M, Valentini E, Zuccarello B. Error and uncertainty analysis of the residual stresses computed by using the hole drilling method. Strain. 2011;**47**: 301-312. DOI: 10.1111/j.1475-1305.2009.00688.x

[27] Peral D, de Vicente J, Porro JA, Ocaña JL. Uncertainty analysis for non-uniform residual stresses determined by the hole drilling strain gage method. Measurement. 2017;**97**:51-63. ISSN: 0263-2241. DOI: 10.1016/j.measurement.2016.11.010

Section 3

Advances in Residual Stress Modelling and Simulation

Chapter 5

Residual Stress Modeling and Deformation Measurement in Laser Metal Deposition Process

Heng Liu and Frank Liou

Abstract

Direct metal deposition (DMD) has become very popular within the space of rapid manufacturing and repair. Its capability of producing fully dense metal parts with complex internal geometries, which could not be easily achieved by traditional manufacturing approaches, has been well demonstrated. However, the DMD process usually comes with high thermal gradients and high heating and cooling rates, leading to residual stresses and the associated deformation, which can have negative effect on product integrity. This paper studies the features of thermal stress and deformation involved in the DMD process by constructing a 3-D, sequentially coupled, thermomechanical, finite element model to predict both the thermal and mechanical behaviors of the DMD process of Stainless Steel 304 (SS 304). A set of experiments were then conducted to validate deformation using a laser displacement sensor. Comparisons between the simulated and experimental results show good agreement. This model can be used to predict the mechanical behavior of products fabricated by the DMD process and to help with the optimization of design and manufacturing parameters.

Keywords: additive manufacturing, residual stress, deformation, finite element analysis, experimental validation

1. Introduction

1.1 Laser-aided direct metal deposition

Laser-aided direct metal deposition (DMD) is an advanced additive manufacturing (AM) technology which can produce fully dense, functional metal parts directly from CAD model. In its operation, the laser beam is focused onto a metallic substrate to create a melt pool, and a powder stream is continuously transported into the melt pool by the powder delivery system. The substrate is attached to a computer numerical control (CNC) multi-axis system, and by moving the substrate according to the desired route pattern, a 2-D layer can be deposited. By building successive layers on top of one another (layer by layer), a 3-D object can be formed. The DMD process has demonstrated its ability in the area of rapid manufacture, repair, and modification of metallic components. Practically, this process is most suitable for components with complex internal geometries that cannot be fabricated by traditional manufacturing methods such as casting.

Furthermore, this process is very cost-effective compared with traditional subtractive manufacturing techniques because it can produce near net shape parts with little or no machining.

1.2 Residual stress and distortion

Residual stresses exist in a part while no external loads were applied. When a part is heated evenly from its previous residual stress-free state, it expands evenly, and no thermal stress is generated. However, when a part is heated unevenly, thermal stress is produced [1].

During DMD processing, highly localized heating and cooling lead to nonuniform thermal expansion and contraction, which further results in a complicated distribution of residual stresses in the heat-affected zone (HAZ) and distortion across the entire structure. These residual stresses could be detrimental—they may cause fractures, promote fatigue, and induce unpredictable buckling during the service of deposited parts; the deformation often is harmful to the dimensional accuracies of structures. Therefore, it becomes critical to predict the two behaviors of materials after the DMD process and to optimize the manufacturing parameters to reduce the residual stresses and deformation.

1.3 Literature review

The thermal behavior of the DMD process has been investigated numerically by many scholars. In [2], a 2-D finite element model was created to simulate the temperature field during the laser cladding process. The results indicated that quasi-steady thermal field could not be reached in a short time. Other scholars have chosen to experimentally investigate thermal behavior. In [3], Griffith et al. employed radiation pyrometers and thermocouples to monitor the thermal signature during laser engineered net shaping (LENS) processing. The results showed that the integrated temperature reheat had a significant effect on the microstructural evolution during the fabrication of hollow H13 tool steel parts. Utilizing a two-wavelength imaging pyrometer, Wang and Felicelli [4] measured the temperature distribution in the melt pool and the area surrounding it during the LENS deposition process. It was found that the maximum temperature in the molten pool is approximately 1600. Only thermal behaviors were investigated in these papers, while no residual stresses were modeled and analyzed.

Some researchers have focused on the modeling and simulation of traditional welding processes, which share many similarities with DMD processes. Using a double-ellipsoid heat source, Gery et al. [5] generated the transient temperature distributions of the welded plates. The results demonstrated that the welding speed, energy input, and heat source distributions had important effects on the shape and boundaries of heat-affected zone. Deng [6] investigated the effects of solid-state phase transformation on the residual stress and distortion caused by welding in low carbon and medium steels. The model discovered that the solid-state phase transformation did not have a noteworthy impact on the final residual stress and the welding deformation in low carbon steel. However, the final residual stresses and the welding deformation appear to be significantly affected by the martensitic transformation in medium carbon steels. Feli et al. [7] analyzed the temperature history and the residual stress field in multi-pass, butt-welded, stainless steel pipes. It was found that in the weld zone and its vicinity, a tensile axial residual stress is produced on the inside surface and compressive axial stress at the outside surface.

Other researchers have attempted to obtain the distribution of residual stress caused by the DMD process through experiments. For example, Moat et al. [8]

measured strain in three directions using a neutron diffraction beam line to calculate the stress in DMD manufactured Waspaloy blocks. They found that large tensile residual stresses exist in the longitudinal direction near the top of the structure. Zheng et al. [9] measured residual stress in PZT thin films fabricated by a pulsed laser using X-ray diffraction. Although experiments can provide relatively accurate results, their flexibility and high cost limit their ability to serve as a general method by which to solve residual stress problems.

In recent years, analyses of the residual stress involved in laser deposition processes using the FE model have been well documented in the literary. In [10], 2-D FE model was built to observe the impact of process parameters on the melt pool size, growth-direction residual stress, and material properties in laser-based deposition processes. They observed that after the deposition was completed and the wall was cooled to room temperature, large tensile stresses exist in the vertical direction at vertical free edges, which is a contrast to the observations in this study. Wang et al. [11] utilized commercial welding software SYSWELD to characterize the residual stress in LENS-deposited AISI 410 stainless steel thin-wall plates. Tensile longitudinal stresses were found near the mid-height, and compressive stresses were found near the top and bottom of the walls. In [12], Kamara et al. investigated the residual stress characteristics of a laser-deposited, multiple-layer wall of Waspaloy on an Inconel 718 substrate. The results indicated that along the length of the wall, residual stresses were almost zero at the bottom and top of the wall. Along the height of the wall, tensile stress with large magnitudes existed at the bottom of the wall while close to the top surface, near stress-free condition, seem to prevail. This matches well with the results presented in this thesis.

1.4 Simulation and experiment approach

Based on the finite element (FE) analysis package ABAQUS, a 3-D, sequentially coupled, thermomechanical model was developed to simulate the transient temperature field, residual stress, and final deformation involved in the DMD process of

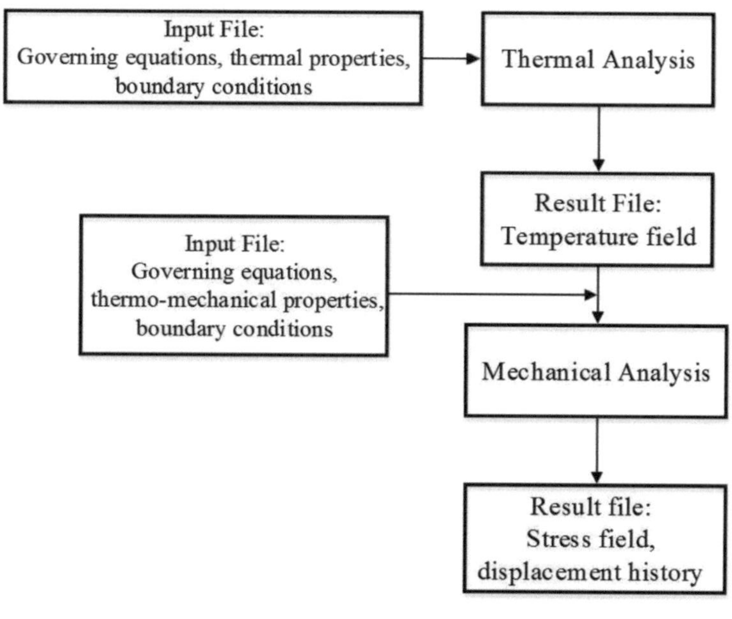

Figure 1.
Flow chart showing the process of numerical modeling.

Stainless Steel 304 (SS 304). The numerical modeling consisted of two major steps shown in **Figure 1**. A transient thermal analysis was firstly carried out to produce the temperature history of the entire workpiece. Then, in the second step, using the temperature field file generated in the previous step as load, a mechanical analysis was carried out to calculate the residual stress and deformation of workpiece.

A laser displacement sensor was used in the experiment to record the vertical deflection of the workpiece resulted from thermal stresses during the deposition process. The accuracy of the numerical model was checked and validated by comparing the experimental results with the simulation results. This validated model can be applied to a multilayer DMD process of stainless steel under different process parameters and can be used for other materials.

2. Thermal analysis

2.1 Governing equations

During DMD processing, the stress and deformation field in the workpiece largely depend on the temperature history; however, the effect of the stress and deformation field on the temperature field is insignificant. Thus, a heat transfer analysis not coupled with mechanical effect is considered.

The transient temperature domain $T(x, y, z, t)$ was attained by solving the heat conduction equation, Eq. (1), in the substrate, along with the initial and boundary conditions discussed in Section 2.2 [13]:

$$\rho C \frac{\partial T}{\partial t} = \frac{\partial}{\partial x}\left(k \frac{\partial T}{\partial x}\right) + \frac{\partial}{\partial y}\left(k \frac{\partial T}{\partial y}\right) + \frac{\partial}{\partial z}\left(k \frac{\partial T}{\partial z}\right) + Q \tag{1}$$

where T is the temperature, ρ is the density, C is the specific heat, k is the heat conductivity, and Q is the internal heat generation per unit volume. All material properties were considered temperature-dependent.

2.2 Initial and boundary conditions

To solve Eq. (1), the initial conditions were set as

$$T(x, y, z, 0) = T_0 \tag{2}$$

$$T(x, y, z, \infty) = T_0 \tag{3}$$

where T_0 is the ambient temperature. In this study, the room temperature of 298 K was used. The boundary conditions, including thermal convection and radiation, are governed by Newton's law of cooling and the Stefan-Boltzmann law, respectively. The heat source parameter, Q in Eq. (1), was considered in the boundary conditions as a surface heat source (moving laser beam). The boundary conditions then could be expressed as [13]

$$K(\Delta T \cdot \boldsymbol{n})|_{\Gamma} = \begin{cases} \left[-h_c(T - T_0) - \varepsilon\sigma\left(T^4 - T_0^4\right)\right]|_{\Gamma} & \Gamma \notin \Lambda \\ \left[Q - h_c(T - T_0) - \varepsilon\sigma\left(T^4 - T_0^4\right)\right]|_{\Gamma} & \Gamma \in \Lambda \end{cases} \tag{4}$$

where k, T, T_0, and Q bear their previous definitions; $\boldsymbol{n}$ is the normal vector of the surface; h_c is the heat convection coefficient; ε is the emissivity which is 0.9; σ is the Stefan-Boltzmann constant which is 5.6704×10^{-8} W/m^2 K^4; Γ denotes the surfaces of the workpiece; and Λ denotes the surface area covered by the laser beam.

2.3 Assumptions and adjustments

Accurate modeling of the thermal process yields highly nonlinear coupled equations, which is time-consuming and expensive to solve. To speed up the solution process and reduce the computational time without sacrificing accuracy, the following assumptions and adjustments were considered.

2.3.1 Energy distribution of the deposition process

In this study, a circular-shaped laser beam with a constant and uniform power density was used. Thus, to match the experiment setup, the heat source parameter Q in Eq. (1) was considered a constant and uniformly distributed surface heat flux defined as

$$Q = \frac{\alpha P}{\pi r^2} \tag{5}$$

where α is the absorption coefficient, P is the power of the continuous laser, and r is the radius of the laser beam. α was set as 0.4 according to the previous experiments conducted, and $r = 1.25$ mm.

2.3.2 Movement of laser beam

The motion of the laser beam was taken into account by updating the position of the beam's center R with time t as follows

$$R = \left[\left(x - \int_{t_0}^{t} udt\right) + \left(y - \int_{t_0}^{t} vdt\right) + \left(z - \int_{t_0}^{t} wdt\right)\right]^{\frac{1}{2}} \tag{6}$$

where x, y, and z are the spatial coordinates and the laser beam centers, u, v, and w, are the continuous velocities the laser beam travels along x-, y-, and z-direction.

In ABAQUS, a user subroutine "DFLUX" [14] was written to simulate the motion of the laser beam.

2.3.3 Powder projection

When modeling, the continuous powder injection process is broken into many small discrete time steps. Using the model change method provided by ABAQUS [14], in each time step, a set of finite elements was added onto the substrate to form deposits along the center line of the substrate. The width of the deposits was assumed to be the same as the diameter of the laser beam, and the thickness of the deposits was calculated from the laser or table travel speed and the powder feed rate. An efficiency of 0.3 was assumed for the power feeding process to account for the powder that did not reach the melt pool.

2.3.4 Modeling the latent heat of fusion

To account for the effect of the latent heat of fusion during the melting and solidification process, the specific heat capacity is modified to generate an equivalent specific heat capacity c_p^* as [15]

$$C_p^*(T) = C_p(T) + \frac{L}{T_m - T_0} \tag{7}$$

where $C_p^*(T)$ is the modified specific heat, $C_p(T)$ is the original temperature-dependent-specific heat, L is the latent heat of fusion, T_m is the melting temperature, and T_0 is the ambient temperature. The values of the latent heat of the fusion, solidus temperature, and liquidus temperature of SS 304 [16] appear in **Table 1**.

2.3.5 Marangoni effect

As discussed in [17], the temperature distribution is significantly impacted by the effect of Marangoni flow, which is caused by the thermocapillary phenomenon. To obtain an accurate thermal field solution, based on the method proposed by [18], an artificial thermal conductivity was used to account for the Marangoni effect:

$$k_m(T) = \begin{cases} k(T) & T \le T_{liq} \\ 2.5 \cdot k(T) & T > T_{liq} \end{cases} \tag{8}$$

where $k_m(T)$ is the modified thermal conductivity, T_{liq} is the liquidus temperature, and T and $k(T)$ maintain their previous definitions.

2.3.6 Combined boundary conditions

The boundary conditions shown in Eq. (4) can be rewritten as

$$K(\Delta T \cdot \boldsymbol{n})|_\Gamma = \begin{cases} [(-h_c - h_r)(T - T_0)]|_\Gamma & \Gamma \notin \Lambda \\ [Q - (-h_c - h_r)(T - T_0)]|_\Gamma & \Gamma \in \Lambda \end{cases} \tag{9}$$

where h_r is the radiation coefficient expressed as

$$h_r = \varepsilon\sigma\left(T^2 + T_0^2\right)(T + T_0) \tag{10}$$

Eq. (9) shows that when the temperature is low, convection is dominant in heat loss and when temperature is high, radiation becomes dominant. As shown in Eq. (10), radiation coefficient is the third-order function of temperature T, which is highly nonlinear. This would greatly increase the computational expense and time. Based on experimental data, an empirical formula combining convective and radiative heat transfer were given by [19] as

$$h = h_c + \varepsilon\sigma\left(T^2 + T_0^2\right)(T + T_0) \approx 2.41 \times 10^{-3}\varepsilon T^{1.61} \tag{11}$$

where h is the combined heat transfer coefficient which is a lower-order function of temperature T compared with h_r. The associated loss in accuracy using this relationship is estimated to be less than 5% [20]. In ABAQUS, a user subroutine "FILM" is written to simulate heat loss.

Latent heat of fusion (J/kg)	Solidus temperature (K)	Liquidus temperature (K)
273,790	1703	1733

Table 1.
Latent heat of fusion for stainless steel 304.

2.4 Finite element modeling

2.4.1 Dimension and parameter

As shown in **Figure 2**, a finite element model for a one-pass, three-layer DMD process was built. The dimension of the substrate under consideration is 50.8 × 12.7 × 3.175 mm (2 × 0.5 × 0.125 inch). Two cases were simulated with different process parameters including laser power, laser travel speed, and powder feed rate. These parameters were chosen according to the criterion that the final geometry of deposits and the total energy absorbed by the specimen be the same in each case. These process parameters are detailed in **Table 2**.

2.4.2 Material properties

Temperature-dependent thermal physical properties of SS 304, including the density, specific heat, thermal conductivity, and latent heat, were used as inputs. The values of these properties appear in **Figure 3**.

2.4.3 Element selection method

Based on the computational accuracy and cost, the type and size of the finite elements used to simulate the substrate and deposits were determined. In transient heat transfer analysis, second-order elements generally produce more accurate results; however, there is a minimum time increment. A simple guideline can be written as [14]

$$\Delta t > \frac{6c}{\rho k} \Delta l^2 \tag{12}$$

where ρ, c, and k have been defined; Δl is the element dimension; and Δt is the time increment. If the time increment Δt is smaller than this value, nonphysical

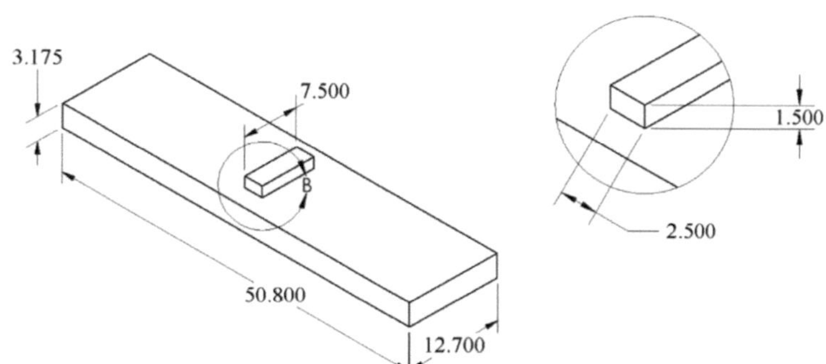

Figure 2.
The dimensions of DMD specimen.

Case number	Laser power (W)	Laser travel speed (mm/min)	Powder feed rate (g/min)
1	607	250	6.3
2	910	375	9.4

Table 2.
DMD process parameters.

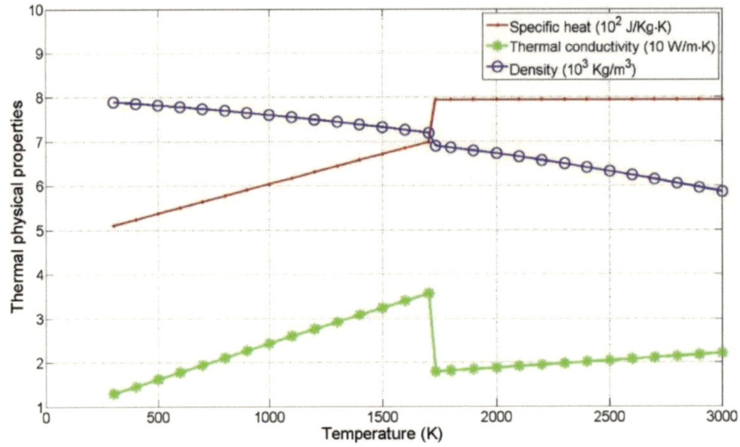

Figure 3.
Temperature-dependent mechanical properties of SS 304.

oscillations may appear in the solution. According to [14], the first-order elements can eliminate such oscillations but may lead to inaccurate solutions [13]. To take both the computational efficiency and accuracy into account, the first-order heat transfer elements, C3D8, with h-version mesh refinement, to refine the mesh by subdividing existing elements into more elements of the same order, were used for the whole domain. Fine meshes were used in zones close to the deposits, and the mesh size progressively increased with the distance from the deposits. In regions more away from the heat-affected zone, coarser meshes were implemented. A total of 14,496 elements and 17,509 nodes was created.

2.4.4 Increment control

In order to obtain reliable results from the mechanical analysis, the maximum nodal temperature change in each increment was set as 5 K, and the time increments were selected automatically by ABAQUS to ensure that this value was not exceeded at any node during any increment of the analysis [14].

3. Mechanical analysis

3.1 Governing equations

The total strain ε_{ij} can be represented generally as

$$\varepsilon_{ij} = \varepsilon_{ij}^{M} + \varepsilon_{ij}^{T} \tag{13}$$

where ε_{ij}^{M} is the strain contributed by the mechanical forces and ε_{ij}^{T} is the strain from thermal loads. Eq. (12) can be decomposed further into five components as [6]

$$\varepsilon_{ij} = \varepsilon_{ij}^{E} + \varepsilon_{ij}^{P} + \varepsilon_{ij}^{T} + \varepsilon_{ij}^{\Delta V} + \varepsilon_{ij}^{Trp} \tag{14}$$

where ε_{ij}^{E} is the elastic strain, ε_{ij}^{P} is the plastic strain, ε_{ij}^{T} is the thermal strain, $\varepsilon_{ij}^{\Delta V}$ is the strain due to the volumetric change in the phase transformation, and ε_{ij}^{Trp} is the strain caused by transformation plasticity. Solid-state phase transformation does not exist in stainless steel [21], so $\varepsilon_{ij}^{\Delta V}$ and ε_{ij}^{Trp} vanish. The total strain vector is then represented as

$$\varepsilon_{ij} = \varepsilon_{ij}^{E} + \varepsilon_{ij}^{P} + \varepsilon_{ij}^{T} \tag{15}$$

The elastic stress-strain relationship is governed by isotropic Hooke's law as

$$\sigma_{ij} = D_{ijkl}\varepsilon_{ij}^{E} \quad (i, j, k, l = 1, 2, 3) \tag{16}$$

where D_{ijkl} is the elastic stiffness tensor calculated from Young's modulus E and Poisson's ratio ν as [11]

$$D_{ijkl} = \frac{E}{1+\nu}\left[\frac{1}{2}\left(\delta_{ik}\delta_{jl} + \delta_{ij}\delta_{kl}\right) + \frac{\nu}{1-2\nu}\delta_{ij}\delta_{kl}\right] \tag{17}$$

where δ_{ij} is the Kronecker delta function defined as

$$\delta_{ij} = \begin{cases} 1 & for \quad i = j \\ 0 & for \quad i \neq j \end{cases} \tag{18}$$

For isotropic elastic solids, Eq. (15) can be simplified as

$$\varepsilon_{ij}^{E} = \frac{1+\nu}{E}\sigma_{ij} - \frac{\nu}{E}\sigma_{kk}\delta_{ij} \tag{19}$$

Thermal strain ε_{ij}^{T} can be calculated from the thermal expansion constitutive equation

$$\varepsilon_{ij}^{T} = \alpha \Delta T \delta_{ij} \tag{20}$$

where α is the thermal expansion coefficient and ΔT is the temperature difference between two different material points. Rate-independent plasticity with the von Mises yield criterion and linear kinematic hardening rule [21] was utilized to model the plastic strain.

Unlike the elastic and thermal strain, no unique relationship exists between the total plastic strain and stress; when a material is subjected to a certain stress state, there exist many possible strain states. So strain increments, instead of the total accumulated strain, were considered when examining the strain-stress relationships. The total strain then was obtained by integrating the strain increments over time t. The plastic strain-stress relationship for isotropic material is governed by the Prandtl-Reuss equation [22]:

$$d\varepsilon_{ij}^{P} = \lambda s_{ij} \tag{21}$$

where $d\varepsilon_{ij}^{P}$ is the plastic strain increment, λ is the plastic multiplier, and s_{ij} is the deviatoric stress tensor defined by

$$s_{ij} = \sigma_{ij} - \frac{1}{3}\sigma_{kk}\delta_{ij} \tag{22}$$

By substituting Eq. (18), Eq. (19), Eq. (20), and Eq. (21) into Eq. (14) and taking the derivative with respect to time, the total strain rate can be described by [2]

$$\dot{\varepsilon}_{ij} = \frac{1+\nu}{E}\dot{\sigma}_{ij} - \frac{\nu}{E}\dot{\sigma}_{kk}\delta_{ij} + \alpha\dot{T}\delta_{ij} + \lambda\left(\sigma_{ij} - \frac{1}{3}\sigma_{kk}\delta_{ij}\right) \tag{23}$$

3.2 Initial and boundary conditions

The temperature history of all the nodes generated in the thermal analysis was imported as a predefined field into the mechanical analysis. The only boundary condition applied to the domain was that the substrate was fixed on one side to prevent rigid body motion. In ABAQUS, the node displacements on the left side of the substrate were set as 0.

3.3 Finite element modeling

3.3.1 Material properties

Temperature-dependent mechanical properties including the thermal expansion coefficient [23], Young's modulus, Poisson's ratio [21], and yield stress [16] were used to model the thermomechanical behavior of SS 304. The values of these properties appear in **Figure 4**.

3.3.2 Element selection

The order of element and integration method used in the mechanical analysis differed from those used in the thermal analysis, while the element dimension and meshing scheme remained unchanged. To ensure the computational accuracy of the residual stress and deformation, second-order elements were utilized in the heat-affected zone, while first-order elements were used in other regions to reduce the computation time. Preventing shear and volumetric locking [14] requires the selection of reduced-integration elements. Therefore, elements "C3D20R" and "C3D8R" in ABAQUS were combined in use to represent the domain.

As shown in **Figure 5**, the 3-D 20-node element used in the mechanical analysis had 12 more nodes than the 3-D 8-node element used in the thermal analysis.

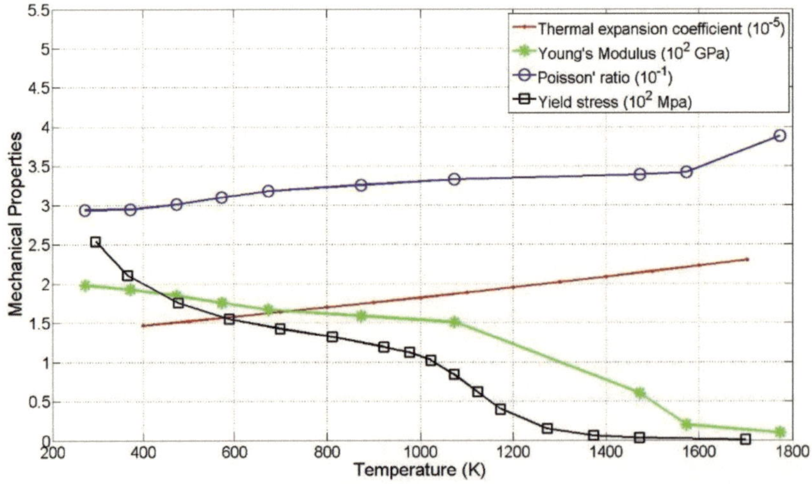

Figure 4.
Temperature-dependent mechanical properties of SS 304.

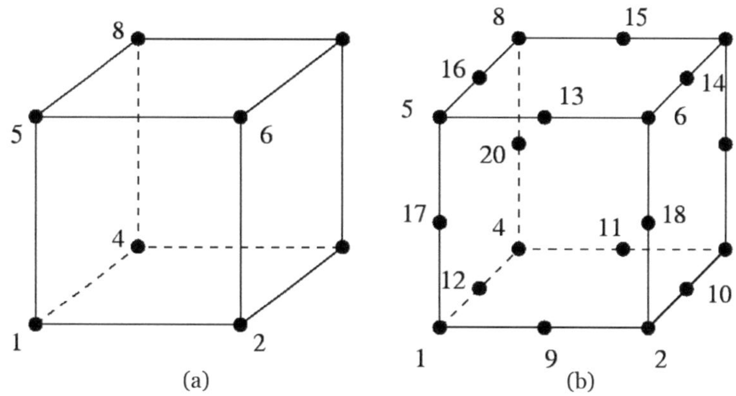

Figure 5.
Elements used in thermal and mechanical analysis. (a) 8-node brick element (b) 20-node brick element.

Therefore, when mapping the temperature data from the thermal analysis to the mechanical analysis, interpolation had to be conducted to obtain the temperature of the 12 extra mid-side nodes (Nodes 9–20 in **Figure 5(b)**).

4. Numerical results and experimental validation

4.1 Temperature

4.1.1 Temperature field

Figure 6 shows the temperature field of the melt pool and surrounding areas from top view at different times in Case 1 (laser power 607 W, laser travel speed 250 mm/min, powder feed rate 6.3 g/min). Laser beam cyclically moves along +z and –z-direction. At $t = 0.9$ s, $t = 2.7$ s, and $t = 4.5$ s, laser beam is located in the center of the substrate. **Figure 7** shows the temperature field and isotherms of the substrate and deposits from the side view at $t = 4.5$ s in Case 1. The peak temperature during the process was around 2350 K, while the lowest temperature was close to room temperature. The big temperature differences and small geometrical dimensions caused very large temperature gradients.

4.1.2 Temperature gradient

The temperature gradient involved in the DMD process was quantitatively analyzed in details. The temperature of nodes along the x' and y' (shown in **Figure 8**) axes in simulation Case 1 at $t = 4.5$ s is shown in **Figure 9**. The x'-direction nodes were selected along the top surface of the substrate (bottom surface of the deposits), while the y'-direction nodes were selected along the height of the deposits. The temperature of the substrate's top surface reached a maximum of 1069 K just below the center of the laser beam and decreased gradually along the x'-direction. In the y'-direction, the temperature of the deposits reached a maximum of 2220 K on the top surface of the deposits and decreased rapidly to1069 K . The slopes of the temperature curves represent the thermal gradients along the x'- and y'-directions. Along x', the temperature gradient reached a maximum of483 K/mm; along y', the maximum temperature gradient occurred near the top surface of the deposits, reaching 1416 K/mm and then decreasing along the negative y'-direction.

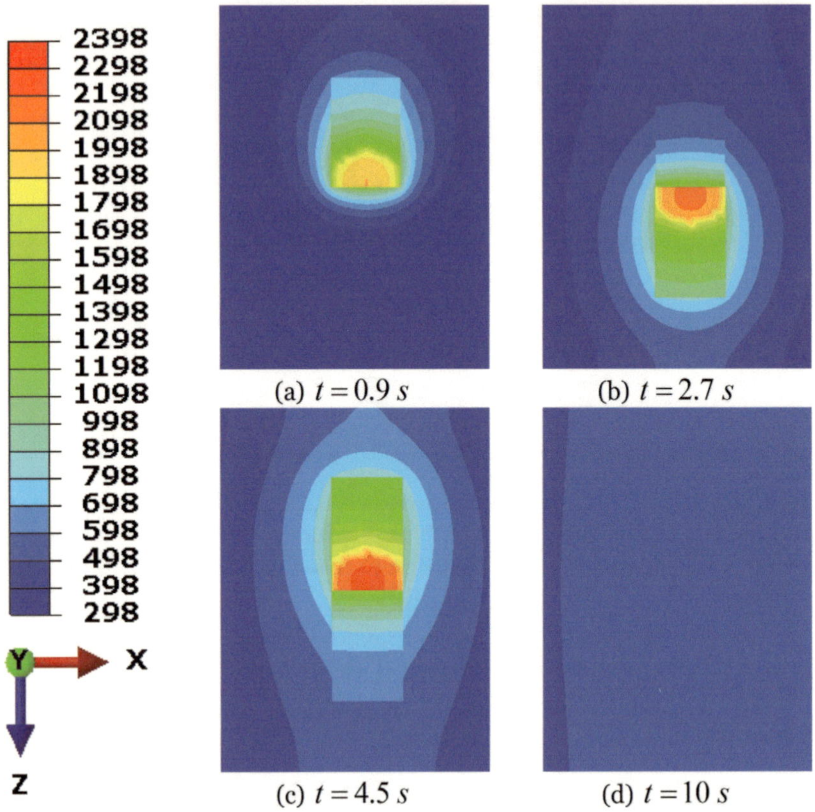

Figure 6.
Contour plots of temperature field of the melt pool and surrounding areas from top view at different times (Case 1).

Figure 7.
Contour plots of temperature field and isotherms of the substrate and deposits from side view at t = 4.5 s (Case 1).

These steep thermal gradients induced large compressive strains within the deposits and substrates [24].

4.2 Residual stress

The nature and magnitude of residual stresses existing in final deposits would affect the integrity of the entire structure. In general conditions, compressive residual stresses are advantageous since they increase the load resistance and

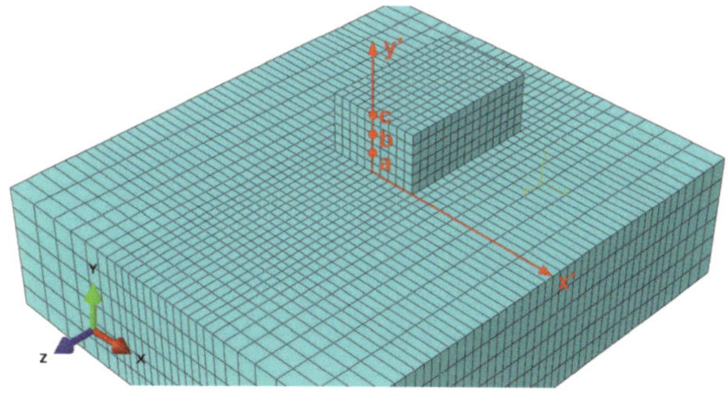

Figure 8.
Location of points within deposition under consideration.

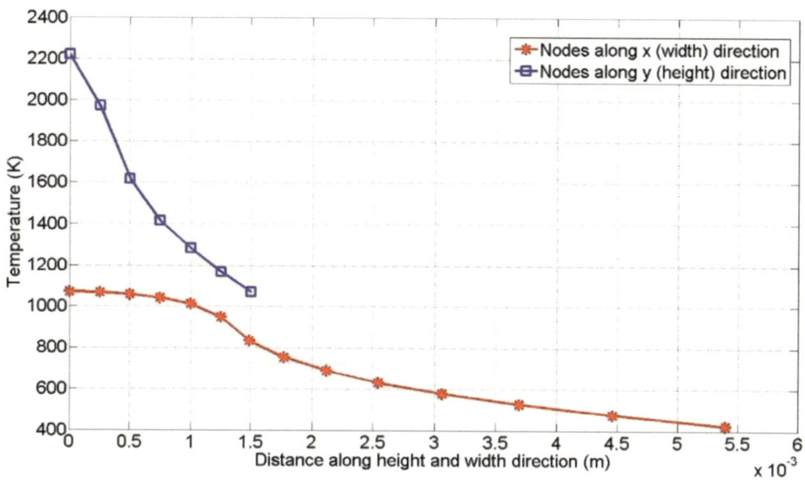

Figure 9.
The temperature of nodes in x- and y-directions in Case 1 at t = 4.5 s.

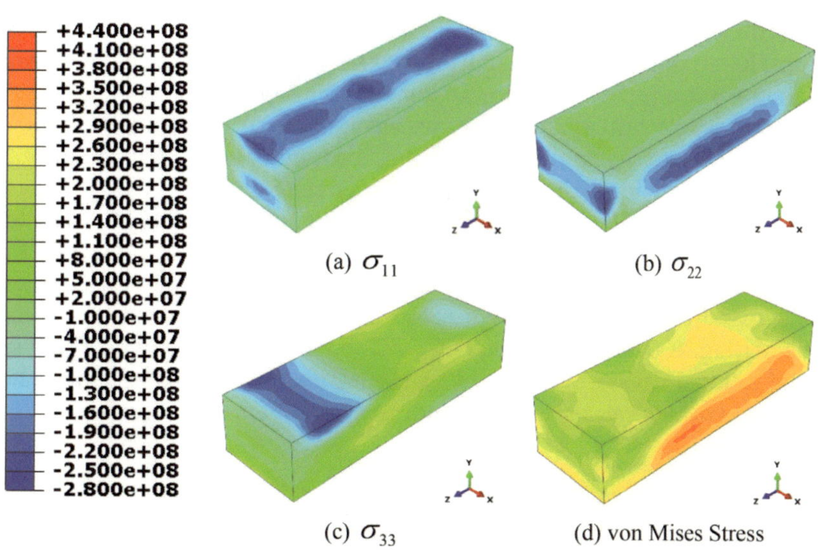

Figure 10.
Contour plots of residual stress field within deposits.

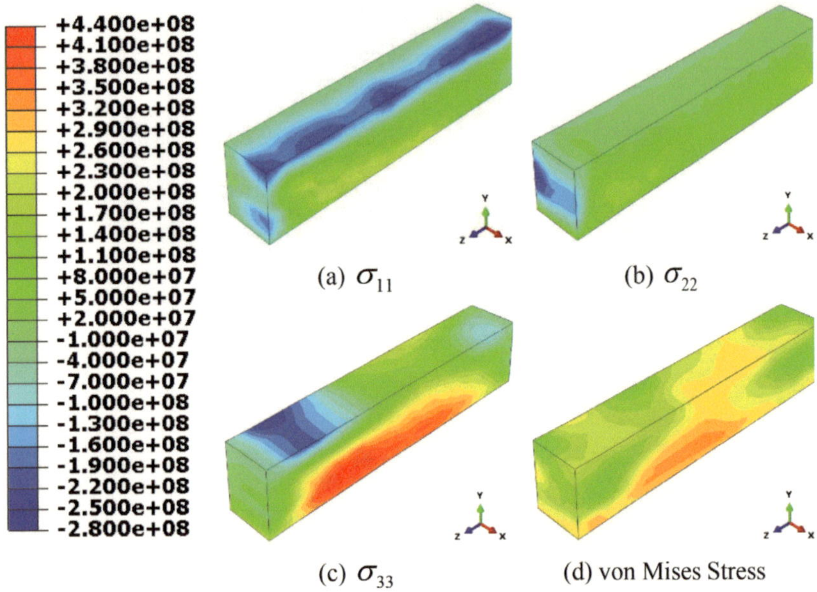

Figure 11.
Contour plots of residual stress field within deposits (y–y cross section).

prevent crack growth, while tensile residual stresses are detrimental that they reduce the load resistance and accelerate crack growth.

The residual stress (in Pa) distribution within the final deposits is shown in **Figures 10** and **11**, where **Figure 10** shows the whole substrate and deposit, while **Figure 11** shows a *y–y* cross section view with half of the deposits hidden to show the internal residual stress. Normal stresses σ_{11}, σ_{22}, and σ_{33} along three spatial directions are shown in **Figures 10** and **11(a)–(c)**, respectively, and the von Mises stress is shown in **Figures 10** and **11(d)**. As the figures indicate, residual stresses in the lower part of the deposits were mostly tensile stresses due to the cooldown phase of the molten layers [24]. After the deposition was finished, the remelted lower part of the deposits began to shrink; this shrinkage was restricted by the underlying material, thus inducing tensile stresses. Compressive residual stresses existed at the top free surface of the deposits, caused by the steep temperature gradient. The expansion of the hotter top layer was inhibited by the underlying material, thus introducing compressive stress at the top surface.

Various experimental methods for measuring residual stress have been developed, such as destructive methods, including incremental hole drilling, layer removal, and crack compliance, and nondestructive methods including X-ray diffraction and neutron diffraction [8, 9, 24–27]. These methods could be used to measure the residual stress directly with relatively good accuracy; however, they usually are not cost-effective or easy to set up. Therefore, instead of measuring the residual stress directly, a flexible indirect method has been developed for residual stress validation. A one-one relationship exists between the deflection of the substrate and residual stress; therefore, by validating the deflection of the substrate, the residual stress results can be validated indirectly.

4.3 Deformation

During the DMD process, the substrate will continuously expand and shrink, finally maintaining a deformed shape (**Figure 12**). In this study, deflection along y was the main deformation under consideration and is shown in **Figure 13**.

4.3.1 Experiment setup

As shown in **Figure 14**, in the experiment, the substrate was clamped at the left end to prevent rigid body motion. Keyence's LK-G5000 series laser displacement sensor shown in **Figure 15** was placed just below the right end of the substrate to record the displacement of the free end along the y-direction with a frequency of 25 Hz during the process. The experimental results appear in **Figure 16**. The entire DMD process was controlled by the "Laser Aided Material Deposition System".

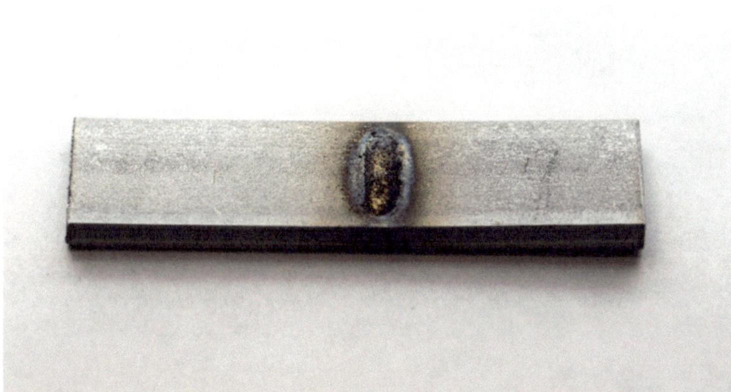

Figure 12.
Final shape of substrate.

Figure 13.
Deflection of substrate along y (m).

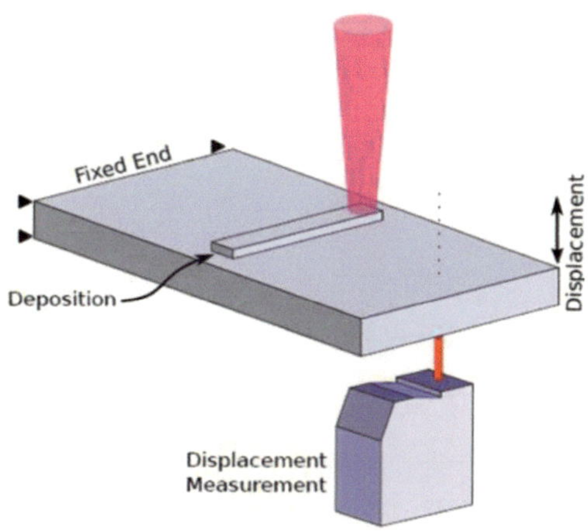

Figure 14.
Experimental setup.

Figure 15.
Displacement sensor.

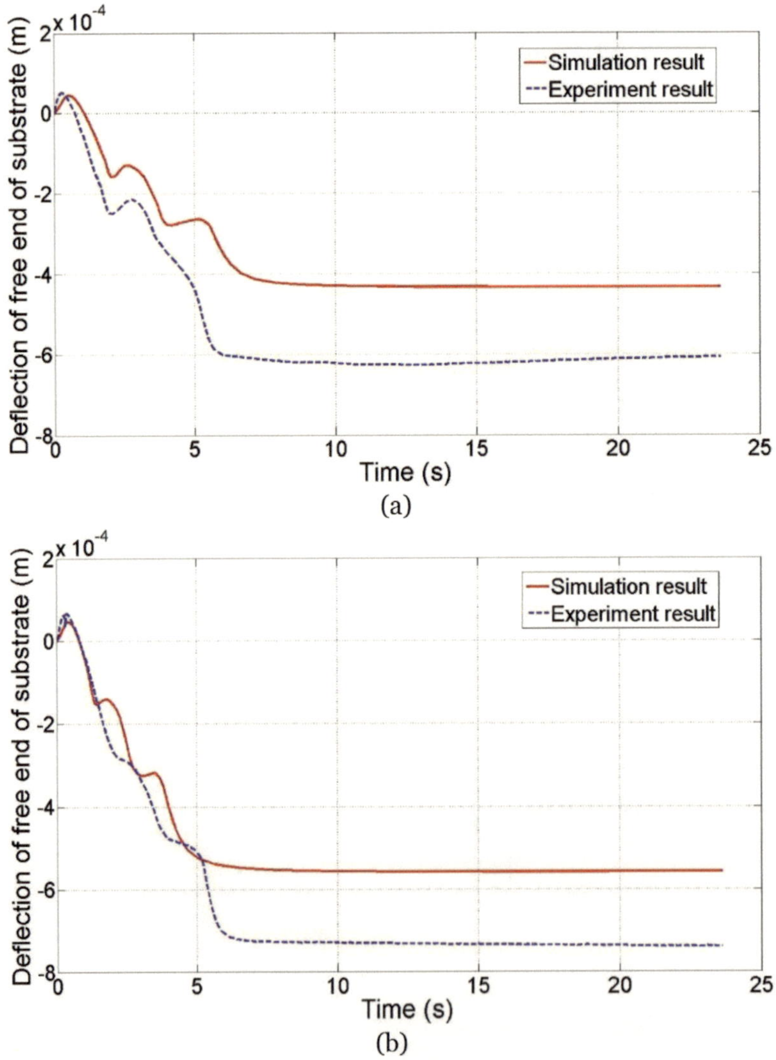

Figure 16.
Simulation and experimental results of substrate deflection. (a) Deflection in case 1 and (b) deflection in case 2.

4.3.2 Experimental and simulation results

Figure 16 illustrates the comparisons of the substrate deflection between the experimental and simulation results for both cases. It is obvious that the simulated deflection matched well with the experimental results. During each deposition layer, the substrate firstly bent down due to thermal expansion on the top surface and then bent up due to thermal shrinkage from the cooling process. After cooling down, the substrate still maintained its distorted shape.

The deflection values from simulation were 28.5 and 24.6% higher than the deflection measured from experiments for Cases 1 and 2, respectively. There are several potential factors that could cause these differences. Firstly, the experimental setup could not perfectly match the setup in the simulation. For example, in the simulation, the laser beam traveled exactly along the centerline of the substrate. While in experiments (**Figure 12**), this cannot be perfectly achieved. These offsets may greatly affect the deflection value since it is very sensitive to the positions of heated zone (where expansion and shrinkage mainly happen) and measuring point. Moreover, the laser displacement sensor did not track the displacement of one particular point on the workpiece. Instead, it sensed the signal reflected by an obstacle, so the positions it tracked were always changing as the substrate continuing to deform. Last but not the least, the simplifications and assumptions considered in both thermal and mechanical analyses could also contribute to the differences between the simulation and experiment. For example, although the substrate material is originally isotropic elastic solid, it may become orthotropic after the DMD process.

It is also worth noting that for Cases 1 and 2, even the total amount of energy applied to the substrate is the same and Case 2 has significantly higher distortion than Case 1, which is caused by higher laser power.

5. Conclusion

To investigate the features of thermal and mechanical behavior of deposited materials involved in the DMD process, a sequentially coupled, thermomechanical finite element model was developed for multilayer DMD process of Stainless Steel 304. The results revealed the characteristics of temperature distribution, residual stress, and deformation within the formed deposits and substrates. A set of experiments were conducted to validate the mechanical effects using a laser displacement sensor. This FEA model can be used to predict the mechanical behavior of products fabricated by the DMD process or similar processes with localized heat sources such as laser sintering, laser cladding, and welding.

Acknowledgements

The support from NASA's grant, under NRA NNX11AI73A, is appreciated. The authors would like to acknowledge William Seufzer and Karen Taminger of NASA Langley Research Center for their mentorship. Support from Missouri S&T's Material Research Center, Manufacturing Engineering program, and Intelligent Systems Center is also greatly appreciated.

Author details

Heng Liu[1] and Frank Liou[2*]

1 Ford Motor Company, Dearborn, MI, United States

2 Missouri University of Science and Technology, Rolla, MO, United States

*Address all correspondence to: liou@mst.edu

References

[1] Masubuchi K. Analysis of Welded Structures: Residual Stresses, Distortion, and their Consequences. Oxford, United Kingdom: Pergamon Press; 1980

[2] Kim J-D, Peng Y. Time-dependent fem simulation of dilution control of laser cladding by adaptive mesh method. KSME International Journal. 2000;**14**(2):177-187

[3] Griffith ML, Schlienger ME, Harwell LD, Oliver MS, Baldwin MD, Ensz MT, et al. Understanding thermal behavior in the lens process. Materials & Design. 1999;**20**(2):107-113

[4] Wang L, Felicelli SD. Analysis of thermal phenomena in LENS™ deposition. Materials Science and Engineering: A. 2006;**435**(625-631)

[5] Gery D, Long H, Maropoulos P. Effects of welding speed, energy input and heat source distribution on temperature variations in butt joint welding. Journal of Materials Processing Technology. 2005;**167**(2):393-401

[6] Deng D. FEM prediction of welding residual stress and distortion in carbon steel considering phase transformation effects. Materials & Design. 2009;**30**(2): 359-366

[7] Feli S, Aalami Aaleagha M, Foroutan M, Borzabadi Farahani E. Finite element simulation of welding sequences effect on residual stresses in multipass butt-welded stainless steel pipes. Journal of Pressure Vessel Technology. 2012;**134**(1):441-451

[8] Moat R, Pinkerton A, Li L, Withers P, Preuss M. Residual stresses in laser direct metal deposited Waspaloy. Materials Science and Engineering: A. 2011;**528**(6):2288-2298

[9] Zheng X, Li J, Zhou Y. X-ray diffraction measurement of residual stress in PZT thin films prepared by pulsed laser deposition. Acta Materialia. 2004;**52**(11):3313-3322

[10] Aggarangsi P, Beuth JL, Griffith ML. Melt pool size and stress control for laser-based deposition near a free edge. In: Solid Freeform Fabrication Proceedings. Austin, TX: University of Texas; 2003. pp. 196-207

[11] Wang L, Felicelli SD, Pratt P. Residual stresses in LENS-deposited AISI 410 stainless steel plates. Materials Science and Engineering: A. 2008; **496**(1):234-241

[12] Kamara A, Marimuthu S, Li L. A numerical investigation into residual stress characteristics in laser deposited multiple layer waspaloy parts. Transactions of the ASME-B-Journal Manufacturing Science Engineering. 2011;**133**(3):031013.1-031013.9

[13] Reddy JN, Gartling DK. The Finite Element Method in Heat Transfer and Fluid Dynamics. Boca Raton, FL: CRC PressI Llc; 2010

[14] Simulia D. ABAQUS 6.11 Analysis User's Manual. 2011. p. 22-2

[15] Toyserkani E, Khajepour A, Corbin S. 3-D finite element modeling of laser cladding by powder injection: Effects of laser pulse shaping on the process. Optics and Lasers in Engineering. 2004;**41**(6):849-867

[16] Ghosh S. Process modeling for solidification microstructure and transient thermal stresses in laser aided DMD process [unpublished doctor of philosophy]. University of Missouri-Rolla. 2006

[17] Alimardani M, Toyserkani E, Huissoon JP. A 3D dynamic numerical approach for temperature and thermal stress distributions in multilayer laser

solid freeform fabrication process. Optics and Lasers in Engineering. 2007; **45**(12):1115-1130

[18] Lampa C, Kaplan AF, Powell J, Magnusson C. An analytical thermodynamic model of laser welding. Journal of Physics D: Applied Physics. 1997;**30**(9):1293-1299

[19] Vinokurov VA. Welding Stresses and Distortion: Determination and Elimination. Boston, Spa, England: British Library Lending Division; 1977

[20] Labudovic M, Hu D, Kovacevic R. A three dimensional model for direct laser metal powder deposition and rapid prototyping. Journal of Materials Science. 2003;**38**(1):35-49

[21] Deng D, Murakawa H. Numerical simulation of temperature field and residual stress in multi-pass welds in stainless steel pipe and comparison with experimental measurements. Computational Materials Science. 2006; **37**(3):269-277

[22] Chakrabarty J. Theory of Plasticity. Boston, MA: Butterworth-Heinemann; 2006

[23] Mercelis P, Kruth J-P. Residual stresses in selective laser sintering and selective laser melting. Rapid Prototyping Journal. 2006;**12**(5):254-265

[24] Casavola C, Campanelli S, Pappalettere C. Experimental analysis of residual stresses in the selective laser melting process. In: Proceedings of 2008 SEM International Conference and Exposition on Experimental and Applied Mechanics; Orlando, Florida, USA. 2008

[25] Tanaka R, Hosokawa A, Ueda T, Furumoto T, Aziz A, Sanusi M. Study on reduction of residual stress induced during rapid tooling process: Influence of heating conditions on residual stress. Key Engineering Materials. 2010;**447** (785-789)

[26] Zaeh MF, Branner G. Investigations on residual stresses and deformations in selective laser melting. Production Engineering. 2010;**4**(1):35-45

[27] Liou FW, Choi J, Landers R, Janardhan V, Balakrishnan S, Agarwal S. Research and development of a hybrid rapid manufacturing process. In: 12th Annual Solid Freeform Fabrication Symposium; Austin, Texas, USA. 2001. pp. 138-145

Chapter 6

Residual Stress Evaluation with Contour Method for Thick Butt Welded Joint

Qingya Zhang, Hong Zhou and Jiangchao Wang

Abstract

Thick plate with high tensile strength steel is increasingly employed for offshore structure fabrication, and welding residual stress is essential for assessment of mechanical performance and fatigue toughness. Therefore, it has been becoming the research issue to evaluate the distribution and magnitude of welding residual stress during butt welding of thick plate. With its advantages, contour method (CM) can be used for longitudinal residual stress evaluation by means of measuring sectional shrinkage after cutting the butt welded joint perpendicular to the welding line. Meanwhile, inverse finite element method (IFEM) code is programmed with C++ language to analyze the measured data to reestablish the welding residual stress. And based on the parallel computation of high-performance server, considering the effect of weld remelting and back-gouging during multi-pass welding process, the welding residual stress is predicted by using efficient thermal elastic plastic finite element method (TEP FEM). Results show that longitudinal residual stress turned from tensile stress in welded vicinity into compressive stress in base metal and the maximum tensile stress is 269 MPa. The computed longitudinal residual stress and welding displacement through TEP FEM are identified with the experimental results. In addition, the back-gouging has an insignificant effect on the residual stress but increases the welding displacement of butt welded joint. The proposed TEP FEM can accurately predict the welding residual stress in welded joint and is also an effective method to control welding displacement.

Keywords: high tensile strength steel, residual stress, CM, IFEM

1. Introduction

With the rapid development of lightweight fabrication in ship industry, thick plate with high tensile strength steel was increasingly employed for marine and offshore structure fabrication [1]. Welding technology with high efficiency is an indispensable process used for ship and offshore fabrication. However, it is inevitable that the residual stress of the welded joint and/or structure was induced by welding. Welding residual stresses can be defined as self-equilibrating stresses in a welded component with the absence of external load [2, 3]. It is known to us all that they can interact with imposed stresses and affect the service performance and structural integrity of welded components. Reliable knowledge of welding residual stress is essential to investigate the root cause of degradation mechanism, carry out

IntechOpen

structural integrity assessment for safety critical components, optimize the design of manufacturing routes, and validate residual stress predictions [4, 5].

Up to now, several qualitative and quantitative techniques have been developed to measure residual stress, which are determined by using the specific elastic constants of material based on the measured strain rather than measured stress. All of them are divided into two categories: destructive techniques and nondestructive techniques [6]. For destructive techniques such as sectioning method, hole drilling method, and CM, the residual stress is measured by its relaxation due to the destruction of the state of the equilibrium of residual stress in a mechanical component. While for nondestructive techniques such as X-ray diffraction method and neutron diffraction method, residual stress is determined based on the relationship between residual stress crystallographic parameters of the material. Among the techniques above, according to the Bueckner's superposition principle, CM can provide a 2D cross-sectional map of residual stress normal to a plane of interest which combines the stress relaxation technology and finite element method [7]. The standard procedure of CM is implemented by the following steps [8]: (1) sample cutting on a plane of interest, (2) contour measurement of the cutting plane or surface, (3) data processing of the measurement results, and (4) residual stress back-calculation by using finite element analysis. CM has found a lot of literatures: Xie [9] estimated the residual stress in thick Ti-6Al-4V alloy welded joint by electron beam welding through finite element method and contour method. Murugan and Narayanan [10] employed both the finite element method and contour method to reveal the residual stress distribution induced by welding in tee joint and found that the experimental results agree well with the predicted stress. What's more, Turski and Edwards [11] efficiently measured the residual stress of 316L stainless steel by utilizing the contour method. Braga et al. [12] studied the welding residual stress profile of butt joints of S355 structural steel through contour method and neutron diffraction. Kainuma [13] investigated the welding residual stress in orthotropic steel decks which had a considerable effect on crack initiation and propagation by using cutting method and magnetostriction method. Woo [14] obtained the two-dimensional maps of the longitudinal residual stress through the thickness of 70-mm thick ferritic steel by using the CM. After that, Woo [15] determined the residual stress in an 80-mm thick ferritic steel by combining the neutron diffraction and CM. In addition to butt welded joint, Liu [16] measured the internal residual stress on inertia friction welding of nickel-based superalloy.

From the reviews above, the CM has obtained a lot of achievements. However, the accuracy of novel embedded cutting contour configuration for thick plate welded joint has not been evaluated. In this paper, the welding residual stress of 30-mm thick plate butt welded joint was investigated combining TEP FEM and CM. What's more, the effect of back-gouging on the residual stress distribution of butt welded joint was discussed.

2. Prediction of welding residual stress by TEP FEM

In this study, the welding residual stress in the butt welded joint through shielded metal arc welding (SMAW) was predicted by TEP FEM, an uncoupled thermal/mechanical formulation procedure, which is mainly composed of two sections: (a) the thermal analysis process and (b) the stress analysis process. Because the former has decisive effect on the latter while the latter has only a small influence on the former, thermal-mechanical behavior during welding is analyzed by using uncoupled thermal/mechanical formulation [17]. During thermal analysis, the 3D welded joint FE model is established firstly using the solid element model according to the dimension of welded joint. The heat source model is considered to be an important aspect, and the double-ellipsoidal volumetric model is employed to simulate the welding transient temperature [18]. The heat flux is determined by the welding current, welding voltage, and welding speed. Besides considering the moving heat source, heat loss due to convection and radiation should also be taken into account in the thermal analysis. And the temperature-dependent thermal properties such as thermal conductivity, specific heat, and density are used.

The mechanical analysis is conducted using the welding temperature histories by thermal analysis as the input load. The same FE model used in thermal analysis is employed here. For mechanical analysis, temperature-dependent mechanical properties such as Poisson's ratio, yield strength, Young's modulus, and linear expansion coefficient are mainly considered. Moreover, the total strain was a summation of the elastic strain, thermal strain, plastic strain, creep strain, and strain induced by phase transformation during welding process, as was shown by Eq. (1). The thermal strain is considered using thermal expansion coefficient, and solid-state phase transformation has insignificant influence on the residual stress and deformation in the mild steel [19], so phase change was neglected in the present study. In addition, because the period with high temperature during the entire thermal cycle was very short (only a few seconds), the creep behavior was also ignored. In addition, the working hardening is neglected in this study since its effect on welding residual stress is not significant for mild steel. The total strain increment at a material point can be expressed as the summation of elastic, plastic, and thermal strains, as was shown by Eq. (2):

$$\varepsilon^{total} = \varepsilon^{elastic} + \varepsilon^{thermal} + \varepsilon^{phase} + \varepsilon^{plastic} + \varepsilon^{creep} \tag{1}$$

$$\varepsilon^{total} = \varepsilon^{elastic} + \varepsilon^{thermal} + \varepsilon^{plastic} \tag{2}$$

To ensure the weld fully penetrated, carbon arc gouging process is usually used to remove root metal. In principle, the root metal was melted by carbon arc and blown away by high pressured gas, increasing the original weld area and changing cross-section weld appearance. In this paper, it assumes that the welding arc plays the same role like carbon arc. Therefore, to investigate the effect of back-gouging on residual stresses in butt welded joint, the root weld bead of main weld was heated again when heating back weld.

3. Experimental procedure

In this section, the experimental procedure was introduced: the first step was to obtain the butt welded joint, and then the out-of-plane welding displacement was measured; the next step was to measure the welding residual stress through CM. Finally, the weld profile of cross-section was obtained.

3.1 Welding work

In this study, the butt welded joint was obtained by SMAW. In detail, the low carbon steel Q235 with the thickness of 30 mm was used as base metal, and the filler metal was J507 welding rod with the diameter of 4 mm. The welding groove was symmetric with the angle of 60°. The detailed dimensions and weld groove are presented in **Figure 1**.

3.2 CM measuring principle

On the basis of the Bueckner's superposition principle, the CM is applied for evaluating residual stress existing in metallic parts or structures. The ideal theoretical implementation of CM used for weld is displayed in **Figure 2**. Step A in **Figure 2** is the undisturbed welded joint and the residual stress that one wishes to determine. In step B, the part was cut in two on the plane x = 0 and the cutting plane deformed due to the near-surface residual stress fully released by the cut; therefore stress in the plane x = 0 was zero. Step C is an analytical step, in which the deformed cut surface is forced back to its original shape; the resulting change in stress is determined. Superimposing the stress state in B with the change in stress from C gives the original residual stress throughout the part. And for arbitrary plane in weld, its residual stress can be determined by the following general expression:

$$\sigma^{A}_{(x,y,z)} = \sigma^{B}_{(x,y,z)} + \sigma^{C}_{(x,y,z)} \tag{3}$$

3.3 CM measurement procedure

3.3.1 Welded specimen cutting

Welded specimen cutting is the first and most important step during the CM procedure as the subsequent steps of surface measurement, data processing, and stress calculation rely on the cutting quality of surface contour. It assumes that the

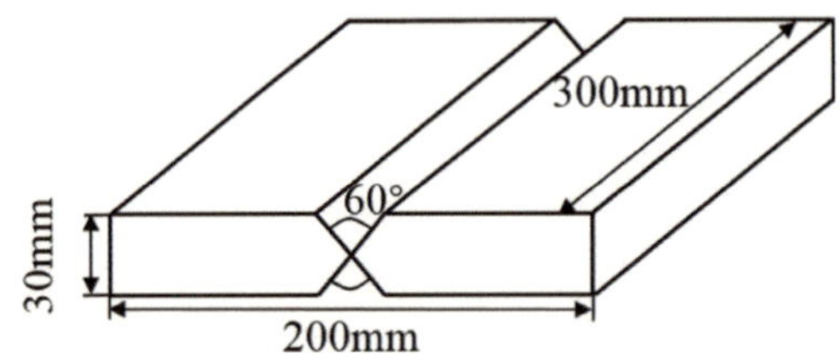

Figure 1.
Weld groove and dimensions.

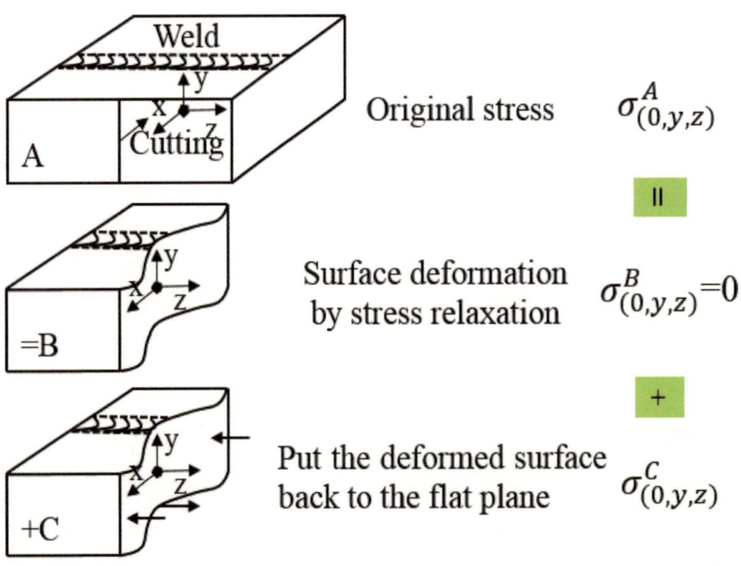

Figure 2.
Schematic diagram of the CM.

cutting process relaxes the residual stress elastically and does not induce any stress into the material. At present, cutting process is implemented by using the wire electrical discharge machining (WEDM) as it can generate a perfectly straight cut and does not remove any further material from the cut surfaces. "Skim cut" setting is used in order to minimize the effect of cutting process on the contour displacement during the relaxation of residual stress. The "skim" mode of WEDM cutting is preferred for contour cuts because of the lower roughness this setting produces. After setting the cutting parameters, the welded specimen is submerged in temperature-controlled deionized water during cutting. In this study, the Sodick AQ400LS with the wire diameter of 0.25 mm is used, and the cutting speed is 0.3 mm/min.

Additionally, in order to minimize the amount of cutting deviates from the original plane, welded specimens should be constrained from moving as stresses are released during the cutting. In conventional (**Figure 3a**) cut configuration, the effects of cutting errors and cutting artifacts cannot be neglected. In Hosseinzadeh's study, plasticity-induced errors in contour measurements can be mitigated by controlling the magnitude of the SIF during cutting. This can be done by choosing an appropriate cutting and restraint strategy. And the SIF is reduced by undertaking an "embedded cut" [20, 21]. Therefore, to mitigate the effects to some extent by restraining the "mode I" opening of cut surfaces during cutting, the embedded cut (**Figure 3b**) is used in this study.

3.3.2 Surface contour measurement

After the welded specimen cutting, the out-of-plane displacement can be implemented through contact measurement and non-contact measurement. Optical machines such as triangulating laser probes, confocal microscopes, et al. are useful for surface measurement. However, handling the large data sets, these systems produced can be problematic, usually requiring some sort of data reduction process. Comparing to non-contact measurement by optical machine, the coordinate measuring machine (CMM) with the uniform measured data is extensively employed for the contour measurement because the regular measured point can be obtained. In this study, the Hexagon micro plus is equipped with a 5-mm diameter touch probe, as is illustrated in **Figure 4**. And each cut surface was sampled with a measurement point spacing of 1 × 1 mm.

3.3.3 Data processing

The procedure of measured data processing is data alignment, data smoothing, averaging of the two data sets, and fitting of the two data sets. The data alignment is

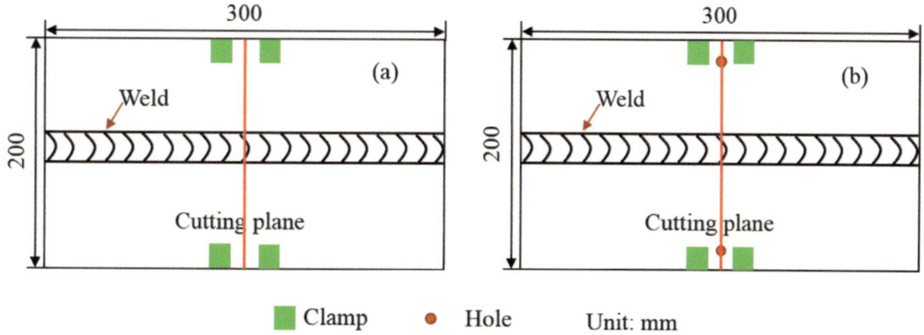

Figure 3.
Schematic drawing of contour cut configuration for welded specimens: (a) conventional contour cut configuration; (b) embedded cut.

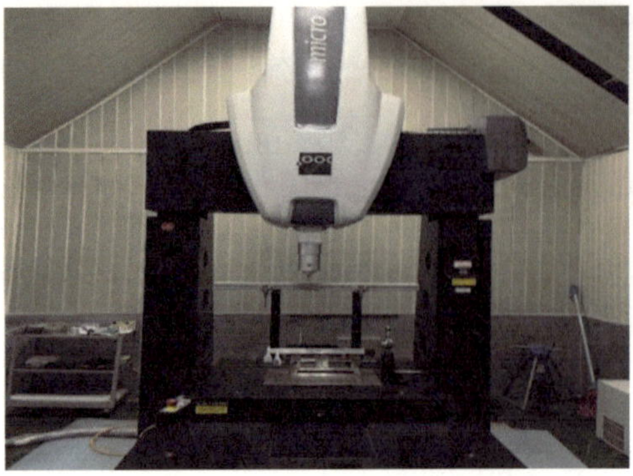

Figure 4.
Hexagon coordinate measuring machine.

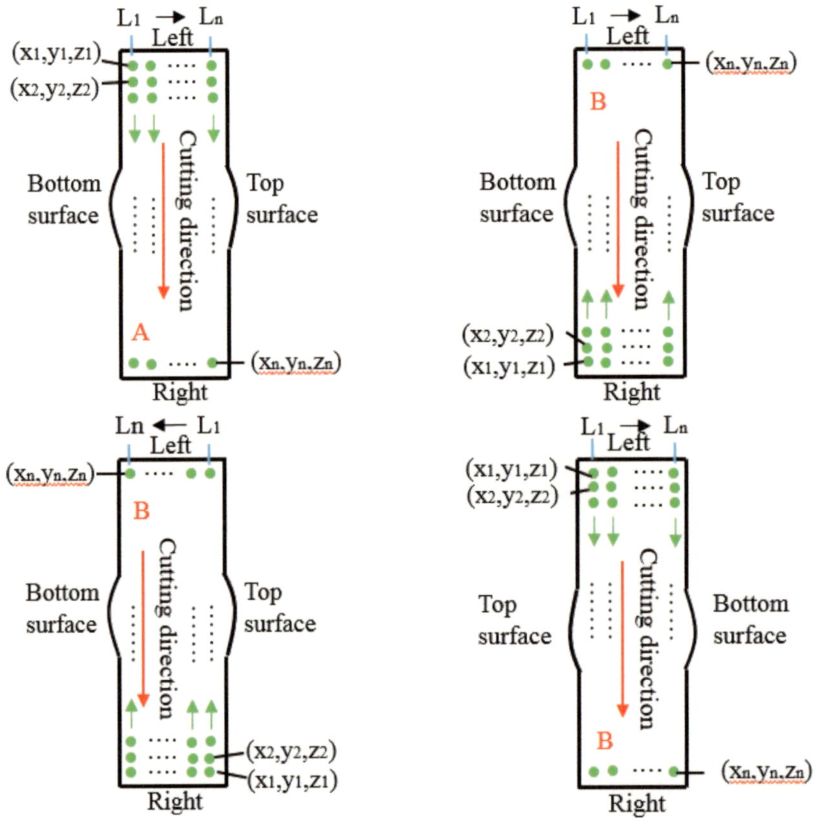

Figure 5.
Data alignment according to the measurement data sequence.

to ensure the raw data on a regular rigid and at the same location. It can be seen from **Figure 5** that during surface contour measurement, several measuring sequence may occur. With the starting point and measuring direction marked, the data in surface B will be the same in surface A by data transposed in Excel or Matlab. Since the presence of noise and outliers is inevitable within the measured raw data, leading to the significant errors in the calculated stresses, the noise and outliers have to be smoothed prior to being applied as the nodal displacement condition in the linear elastic finite element analysis. The averaging of the two data sets can cancel out the shear stress effects and other errors. Fitting of the two data sets can be conducted by spline, Fourier, or polynomial. In this study, the "sgolay" method was used to remove noise and outliers and also fit the measured data through Matlab: the processed data was displayed in **Figure 6**.

3.3.4 Stress calculation

In the final step of CM, the residual stress normal to a plane of interest can be obtained by linear elastic finite element analysis. In this step, the negative processed

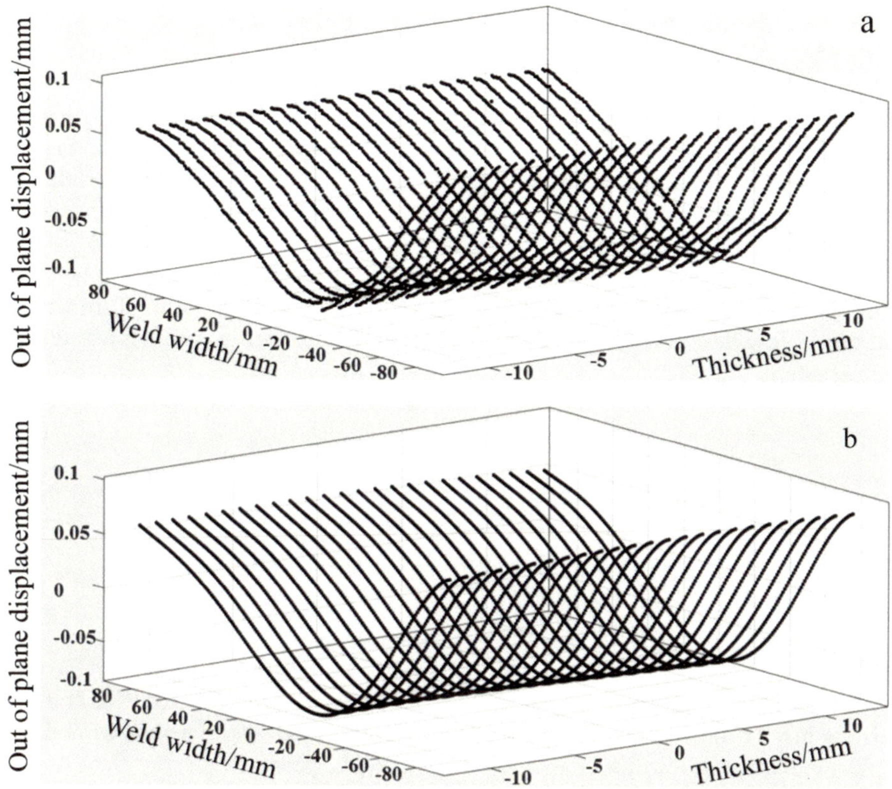

Figure 6.
The processed data of contour: (a) raw data after average; (b) after processing.

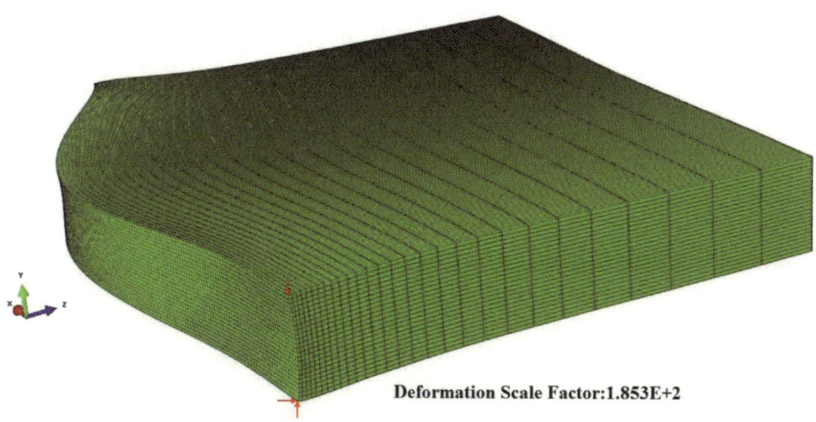

Figure 7.
Finite element model for linear elastic analysis.

data is applied as the nodal displacement condition of the finite element model which is often created according to the half of the whole welded specimen by using ABAQUS or ANSYS code. **Figure 7** displayed the finite element model of half welded joint, and red arrows representing boundary condition prevented the movement of rigid body. The inverse of the measured contour has been applied to the cut surface (deformation in this figure has been magnified ×185). A fine element size (1 mm) is that faces the cutting surface in addition to a fine mesh density around the cutting plane are used in the present FE model.

4. IFEM

According to the concept of CM, the residual stress existing in welded specimens can be calculated through the measured out-of-plane displacement normal to a plane of interest. Several assumptions are made in the linear elastic finite element analysis: (1) small displacements or deformations of the welded components or structures, (2) linear elastic behavior during the material removal, and (3) unchanged boundary conditions during loading process. Therefore, it can be defined as an inverse finite element method (IFEM). In general, a set of linear equation below describes the welded components or structures in the linear elastic finite element:

$$[K][u] = [F] \tag{4}$$

$$[K] = \frac{E}{(1-2\upsilon)(1+\upsilon)}\begin{bmatrix} \frac{\partial^2(1-\upsilon)}{\partial x^2} + \left(\frac{\partial^2}{\partial y^2} + \frac{\partial^2}{\partial z^2}\right)\frac{(1-2\upsilon)}{2} & \frac{\partial^2}{2\partial x \partial y} & \frac{\partial^2}{2\partial x \partial z} \\ \frac{\partial^2}{2\partial x \partial y} & \frac{\partial^2(1-\upsilon)}{\partial y^2} + \left(\frac{\partial^2}{\partial x^2} + \frac{\partial^2}{\partial z^2}\right)\frac{(1-2\upsilon)}{2} & \frac{\partial^2}{2\partial y \partial z} \\ \frac{\partial^2}{2\partial x \partial z} & \frac{\partial^2}{2\partial y \partial z} & \frac{\partial^2(1-\upsilon)}{\partial z^2} + \left(\frac{\partial^2}{\partial x^2} + \frac{\partial^2}{\partial y^2}\right)\frac{(1-2\upsilon)}{2} \end{bmatrix}$$

where E is the Young's modulus, 210 GPa, υ is Poisson's ratio, 0.3, $[K]$is the stiffness matrix of the welded component or structure, $[u]$ is the matrix of nodal displacements, and $[F]$ is the matrix of external nodal force.

5. Results and discussion

The experimental results and simulation results were discussed in section. Based on the reliable welding temperature simulation, the longitudinal residual stress through TEP FEM and CM was analyzed. After that the transverse welding residual stress through TEP FEM was obtained. The comparison analysis of welding displacement was conducted to further validate the efficiency and accuracy of the proposed TEP FEM.

5.1 Maximum welding temperature distribution

Figure 8 displays two FEM models of the butt welded joint, which contain 13,702 nodes, 12,300 elements and 22,072 nodes, 20,400 elements, respectively, and the mechanical boundary constraint conditions are displayed in **Figure 8** by the arrows. It can be evidently seen from the picture that there are a large amount of nodes in the vicinity of weld and mesh toward the plate edge gradually reduced. The way of mesh arrangement is useful for improving computation efficiency and

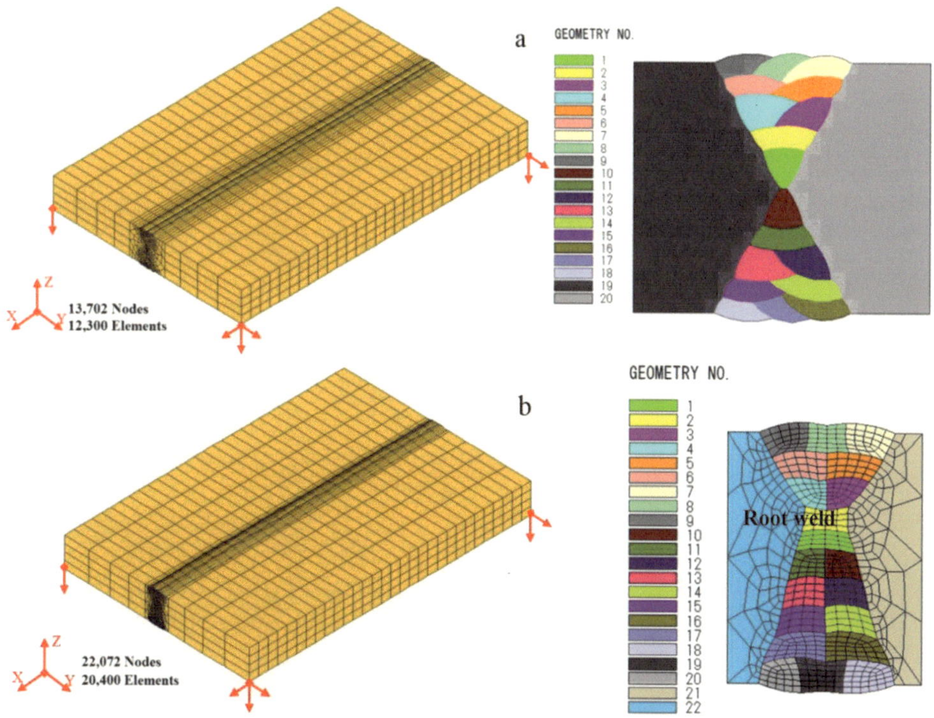

Figure 8.
Simulation model and weld sequence of butt welded joint: (a) without considering back-gouging; (b) with considering back-gouging.

ensuring computation accuracy when carrying out transient welding temperature nonlinear heat transfer computation.

5.1.1 Root weld

In principle, plastic strain is the generation source of welding residual stress, which is mainly determined by the maximum welding temperature and restraint condition. Here, the restraint condition is not considered, and the maximum welding temperature distribution is used to describe the features of welding temperature field of the butt welded joint. During the thermal analysis of the butt welded joint, the initial temperature is assumed to be 20°C. To simulate the heat input of moving welding arc during welding, the volumetric heat source with uniform density distribution is employed. For this heat source, the welding arc energy is dependent on the welding current, voltage, speed, and arc efficiency. When the welding process is SMAW, the arc heat efficiency is 0.6 and the welding arc length is 30 mm. And the heat source volume denoted the considered weld pool volume and can be obtained by calculating the volume fraction of the elements in currently being welded zone. The nonlinear isotropic Fourier heat flux was also employed for heat conduction. The temperature-dependent thermal properties such as thermal conductivity, specific heat, and density are considered. **Figure 9** presents the fusion zone in middle cross-section of the butt welded joint with or without considering back-gouging. The simulated fusion zone is slightly greater than the X-groove area, while the maximum temperature is approximately 2300°C. Comparing to the macrostructure of the butt welded joint, the fusion zone with considering back-gouging agrees better than that without considering back-gouging. Therefore,

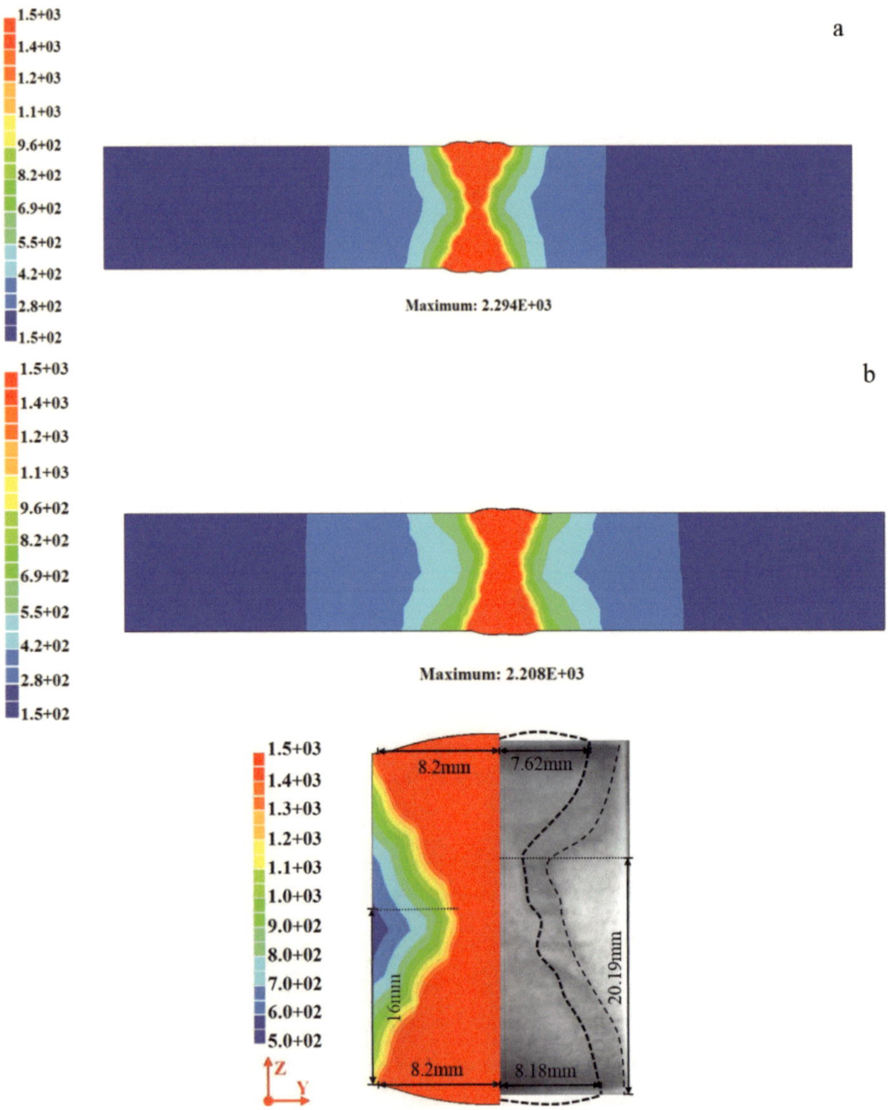

Figure 9.
Fusion zone of butt welded joint: (a) without considering back-gouging; (b) with considering back-gouging.

it can be concluded that the fusion area predicted by FEM with considering back-gouging is reasonable.

5.2 Welding residual stress distribution

Figure 10a–c displays the 2D mapping of longitudinal welding residual stresses in the middle cross-section computed by the numerical simulation and IFEM, respectively. In detail, **Figure 10a** presents the longitudinal residual stresses without considering back-gouging, **Figure 10b** shows the longitudinal residual stresses with considering back-gouging, and **Figure 10c** illustrates the longitudinal residual stresses by IFEM. It can be seen from the figure that both the longitudinal residual stresses with or without considering back-gouging matched with the measured contour. But, the peak value of longitudinal residual stress with considering back-gouging is higher than that without considering back-gouging, which is almost equal to the maximum measured value. Moreover, back-gouging process can reduce

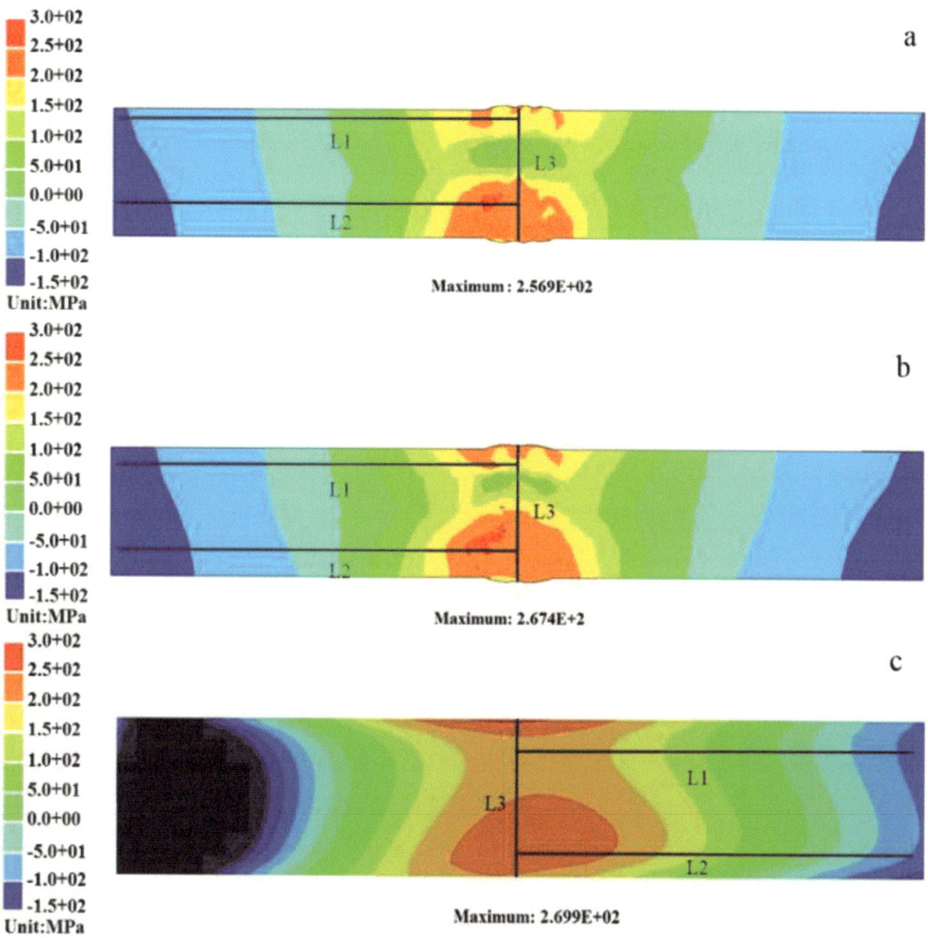

Figure 10.
2D mapping of longitudinal welding residual stress distribution of middle cross-section: (a) without considering back-gouging; (b) with considering back-gouging; (c) IFEM.

the tensile stress area of main weld. In addition, **Figure 10c** shows some irregularities at the edges of the cut surfaces; these irregularities may be produced due to wire entrance and exit during the specimen cutting.

Figure 11a–c quantitatively compares the longitudinal welding residual stress distributions in the middle cross-section along L1, L2, and L3 computed by TEP FE and the corresponding measurements, respectively. It can be seen from the **Figure 11a, b** that the computed longitudinal residual stress distributions along L1 and L2 agree well with the measured stress distribution. It can be seen from **Figure 11c** that the through-thickness longitudinal stresses distribution in center weld was obviously smaller than that in cap welds, which can be increased by back-gouging process. Therefore, it can be found that the peak longitudinal stresses were in cap welds and the longitudinal stresses in center weld can be increased by back-gouging process.

Figure 12a, b displays the features of transverse residual stresses in butt welded joint with or without considering back-gouging, respectively. It can be seen from the picture that the tensile stress is almost constant, while the compressive stress considering back-gouging is greater than that without considering back-gouging. And the middle cross-section mappings show that the signal of transverse residual stresses of local root weld and its distribution are changed obviously. Finally, the

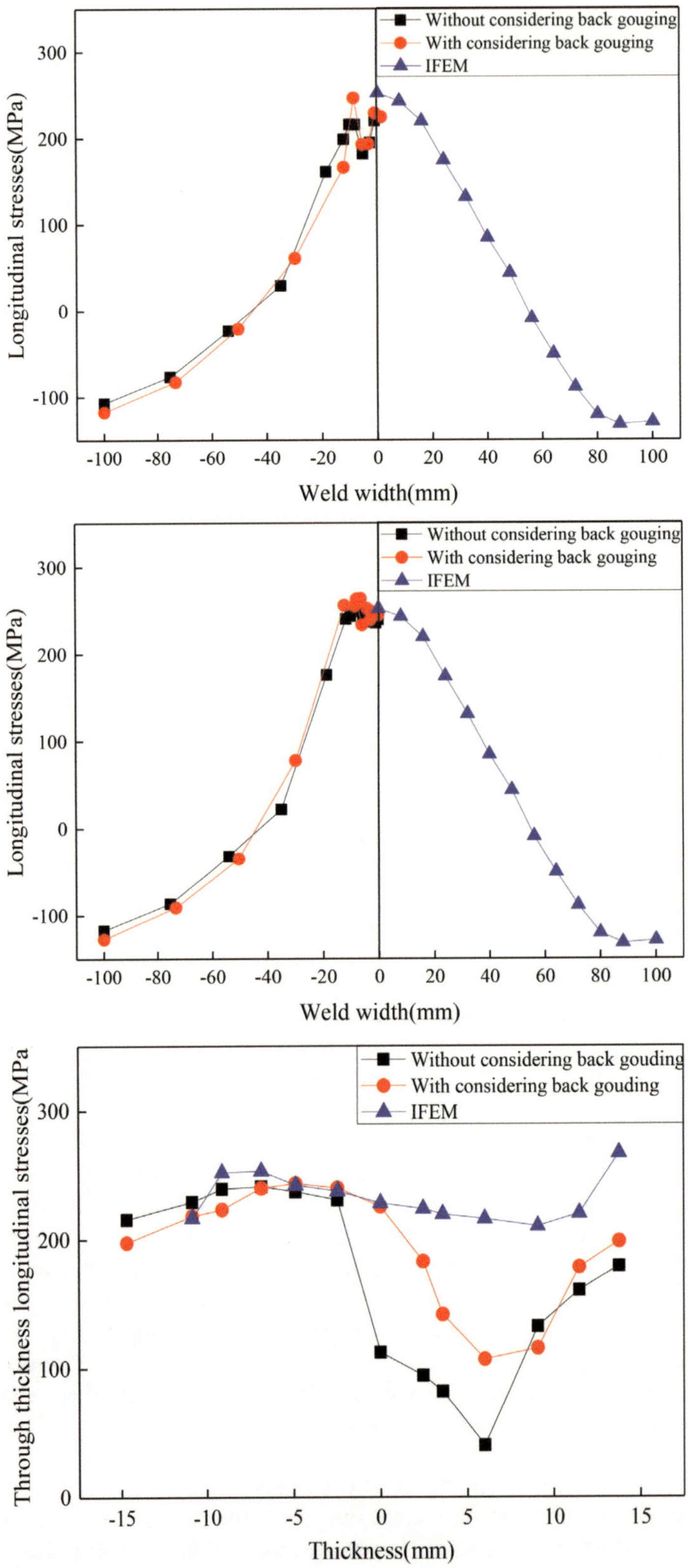

Figure 11.
Longitudinal residual stress distributions along L1, L2, and L3.

locations of the maximum residual stress and through-thickness spatial distributions provide an indication of the significance of crack initiation and the integrity of the welded components.

5.3 Welding displacement

Figure 13 showed the contour of z-direction welding displacement through TEP FEM and CMM measurement. The maximum z-direction welding displacement by TEP FEM was 0.9363 mm without considering back-gouging and 1.5874 with considering back-gouging, respectively. The maximum z-direction displacement by

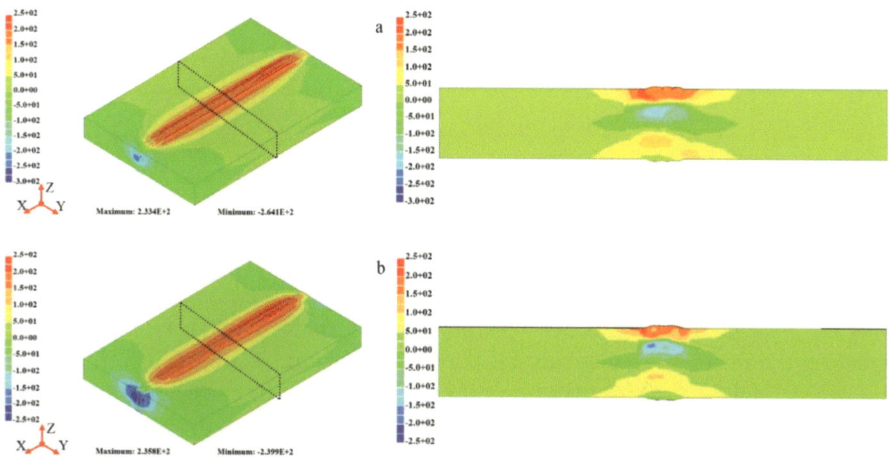

Figure 12.
Transverse welding residual stress distribution: (a) without considering back-gouging; (b) with considering back-gouging.

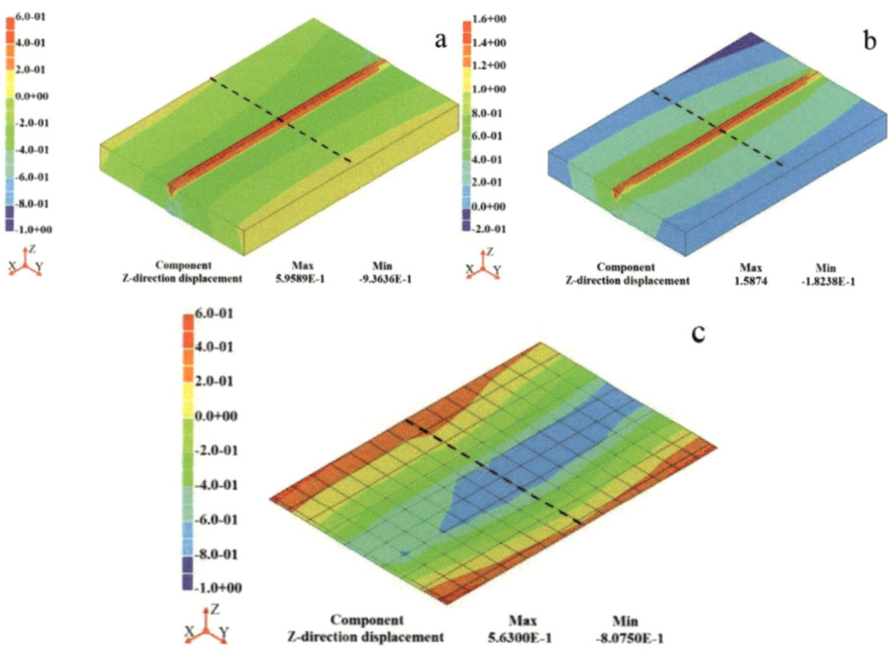

Figure 13.
Contour of z-direction welding displacement: (a) without considering back-gouging, (b) with considering back-gouging, and (c) experimental measurement.

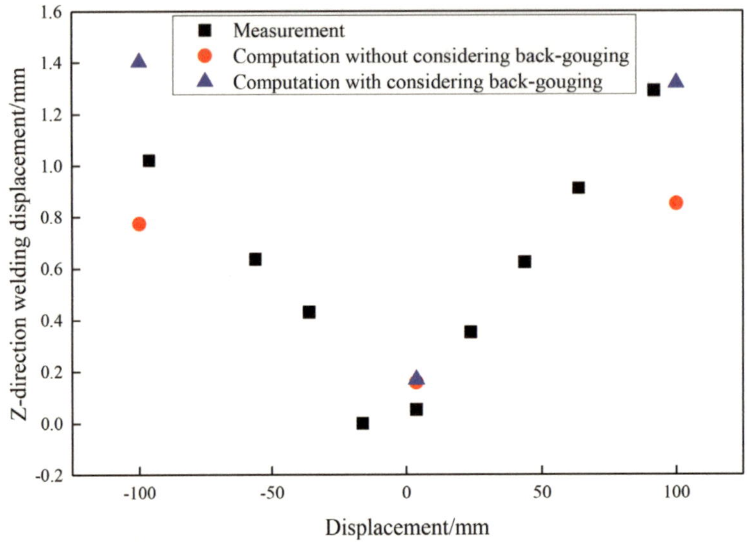

Figure 14.
Comparison of z-direction welding displacement.

CMM measurement was 0.80750. The quantitative analysis was carried out by extracting the mid-length displacement, as was displayed in **Figure 14**. Taking the weld center as the deformation base point, it can be seen that the computation of z-direction welding displacement considering back-gouging had a better agreement with the measurement. It is analyzed that the deformation after main welding was mitigated by the back-gouging process. Therefore, it can be concluded that the residual stress distribution and magnitude obtained by the efficient TEP FEM considering the back-gouging process were accurate and effective.

6. Conclusions

In this paper, the welding-induced residual stress in thick plate butt welded joint is evaluated through numerical simulation and CM. To validate the computed residual stresses, CM measurement is conducted. The results of the CM measurement are compared with the residual stresses computed by TEP FE. Based on the simulation and measured results, the following conclusions can be drawn:

1. Both the computed longitudinal residual stresses along weld width and through-thickness residual stresses are in a good agreement with the CM measured results.

2. Considering the back-gouging, the computed welding displacement through TEP FEM is identified with the measured result.

3. Through the comparison analysis of measurement and computation, the back-gouging has an insignificant effect on welding residual stress distribution and magnitude but can obviously increase the welding displacement.

Acknowledgements

The authors gratefully appreciate the National Natural Science Foundation of China (Grant No. 51609091), Priority Academic Program Development of Jiangsu

Higher Education Institutions, Collegiate Natural Science Fund of Jiangsu Province (14KJA570001), and High-tech Ship Scientific Research of Ministry of Industry and Information Technology of the People's Republic of China in 2017.

Author details

Qingya Zhang[1,2], Hong Zhou[1] and Jiangchao Wang[2*]

1 School of Naval Architecture and Ocean Engineering, Jiangsu University of Science and Technology, Zhenjiang, China

2 School of Naval Architecture and Ocean Engineering, Huazhong University of Science and Technology, Wuhan, China

*Address all correspondence to: wangjiangchaocn@gmail.com

References

[1] Wang JC, Zhao HQ, Zou JS, Zhou H, Wu ZF, Du SD. Welding distortion prediction with elastic FE analysis and mitigation practice in fabrication of cantilever beam component of jack-up drilling rig. Ocean Engineering. 2017; **130**:25-39

[2] Withers PJ, Bhadeshia HKDH. Residual stress. Part 1—Measurement techniques. Materials Science and Technology. 2001;**17**:355-365

[3] Withers PJ, Bhadeshia HKDH. Residual stress. Part 2—Nature and origins. Materials Science and Technology. 2001b;**17**:366-375

[4] Hosseinzadeh F, Kowal J, Bouchard PJ. Towards good practice guidelines for the contour method of residual stress measurement. The Journal of Engineering. 2014;**8**: 453-468

[5] Sun YL, Roy MJ, Vasileiou AN, Smith MC, Francis JA, Hosseinzadeh F. Evaluation of errors associated with cutting-induced plasticity in residual stress measurements using the contour method. Experimental Mechanics. 2017; **57**:719-734

[6] Withers PJ, Turski M, Edwards L, Bouchard PJ, Buttle DJ. Recent advances in residual stress measurement. International Journal of Pressure Vessels and Piping. 2008;**85**:118-127

[7] Prime MB. Cross-sectional mapping of residual stresses by measuring the surface contour after a cut. Journal of Engineering Materials and Technology. 2001;**123**:162-168

[8] Prime MB, DeWald AT. The contour method. In: Schajer GS, editor. Practical Residual Stress Measurement Methods. Hoboken, New Jersey, USA: John Wiley& Sons, Ltd; 2013. pp. 109-138

[9] Xie P, Zhao HY, Wu B, Gong SL. Using finite element and contour method to evaluate residual stress in thick Ti-6Al-4V alloy welded by electron beam welding. Acta Metallurgica Sinica (English Letters). 2015;**28**(7):922-930

[10] Murugan N, Narayanan R. Finite element simulation of residual stresses and their measurement by contour method. Materials & Design. 2009; **30**(6):2067-2071

[11] Turski M, Edwards L. Residual stress measurement of a 316l stainless steel bead-on-plate specimen utilising the contour method. International Journal of Pressure Vessels and Piping. 2009;**86**(1):126-151

[12] Braga DFO, Coules HE, Pirling T, Richter-Trummer V, Colegrove P, de Castro PMST. Assessment of residual stress of welded structural steel plates with or without post weld rolling using the contour method and neutron diffraction. Journal of Materials Processing Technology. 2013;**213**(12): 2323-2328

[13] Kainuma S, Jeong Y-S, Yang M, Inokuchi S. Welding residual stress in roots between deck plate and U-rib in orthotropic steel decks. Measurement. 2016;**92**:475-482

[14] Woo W, An GB, Kingston EJ, DeWald AT, Smith DJ, Hill MR. Through-thickness distributions of residual stresses in two extreme heat-input thick welds: A neutron diffraction, contour method and deep hole drilling study. Acta Materialia. 2013;**61**: 3564-3574

[15] Woo W, An GB, Em VT, De Wald AT, Hill MR. Through-thickness distributions of residual stresses in an 80 mm thick weld using

neutron diffraction and contour method. Journal of Materials Science. 2015;**50**:784-793

[16] Liu C, Zhu H-Y, Dong C-L. Internal residual stress measurement on inertia friction welding of nickel-based super alloy. Science and Technology of Welding and Joining. 2014;**19**(5): 408-415

[17] Radaj D. Heat Effects of Welding: Temperature Field, Residual Stress and Distortion. Berlin/Heidelberg: Springer Verlag Publishing; 1992

[18] Goldak JA, Akhlaghi M. Computational Welding Mechanics. Ottawa: Springer; 2005. pp. 71-115

[19] Yang YP, Dong P. Buckling distortion and mitigation techniques for thin section structures. Journal of Materials Engineering and Performance. 2012;**21**(2):153-160

[20] Hosseinzadeh F, Traore Y, Bouchard PJ, Muránsky O. Mitigating cutting-induced plasticity in the contour method. Part 1: Experimental. International Journal of Solids and Structures. 2016;**94–95**:247-253

[21] Muransky O, Hosseinzadeh F, Hamelin CJ, Traore Y, Bendeich PJ. Investigating optimal cutting configurations for the contour method of weld residual stress measurement. International Journal of Pressure Vessels and Piping. 2018;**164**:55-67

Chapter 7

Durability Assessment Considering Residual Stress

Byung-Ik Kang

Abstract

The relative movement between the cylinder block and the piston may cause a seizure, and this seizure problem can be solved by pressing the bushing into the cylinder block. However, if the rates of shrink fit are insufficient, the copper bushing will be detached from the cylinder block, and if the rates of shrink fit are excessive, a residual stress higher than yield stress will be generated and adversely affect the durability of bushing and the cylinder block. Therefore, in this study, a clear and quantitative rate of shrink fit is presented to solve the above-mentioned problems, and the durability assessment process of mechanical parts where residual stress occurs is performed. Through this study, analytical technique that can predict the residual stress clearly according to the rates of shrink fit was obtained and durability assessment was completed based on the quantitative residual stress according to the rates of shrink fit.

Keywords: durability assessment, cylinder block, copper bushing, shrink fit, seizure

1. Introduction

The cylinder block of the main pump is an important part of the hydraulic system and generates high-pressure hydraulic fluid through relative motion with pistons. At this time, there may occur a seizure problem [1] in which the surface of each component is adhered. And the problem may become a serious problem leading to the shutdown of the entire machine. Therefore, in order to prevent this shutdown, the cylinder block may be fabricated by copper or the copper bushing may be pressed into the cylinder block. Among these two methods, manufacturers mainly use the method of pressing the copper bushing into the cylinder block to reduce the production cost. However, if the proper rate of shrink fit can not be found when pressing the copper bushing, it may lead to disconnection of the bushing or breakage of the cylinder block in the use environment of the machine, resulting in the shutdown of the entire machine, such as seizure. Therefore, in this study, the method of setting the rate of shrink fit analytically was confirmed, and the durability assessment method and the process of the cylinder block considering the residual stress generated after the pressing copper bushing were constructed.

2. Antiseizure cylinder block

A typical cylinder block manufacturing method is shown on the left of **Figure 1**. At this time, the production process of each process does not generate residual

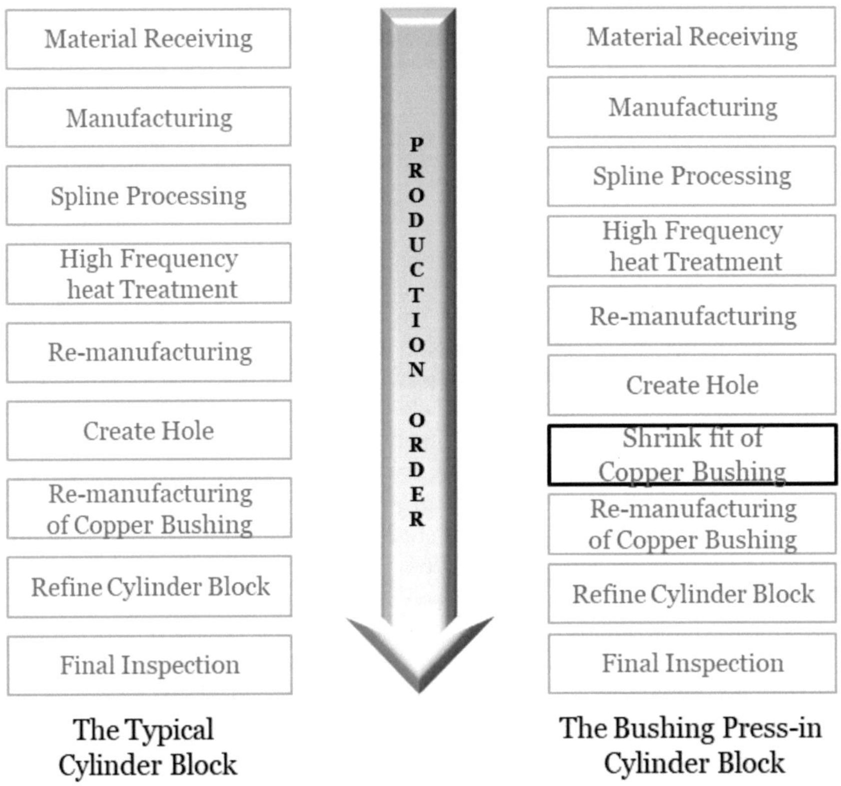

Figure 1.
Comparison of production processes.

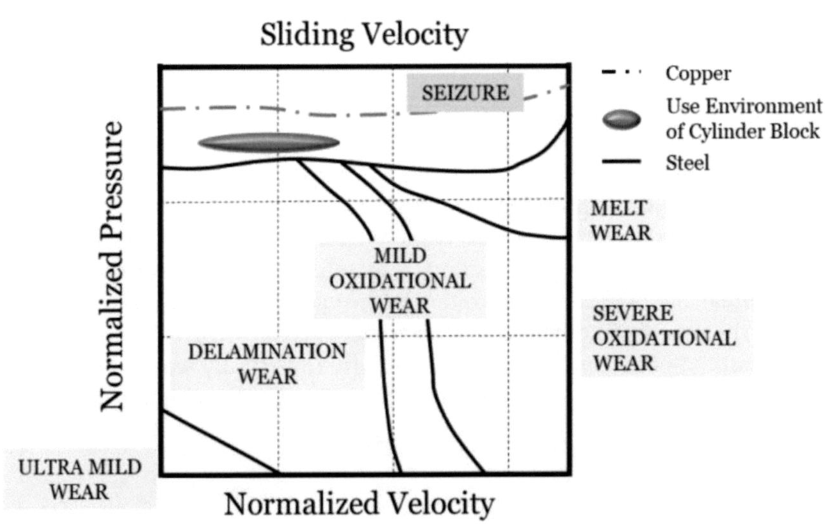

Figure 2.
Wear-mechanism map.

stress enough to have a large effect on the final product [2]. However, when a copper bushing is inserted into the cylinder block to prevent the seizure between the cylinder block and the piston (on right of **Figure 1**), and residual stresses corresponding to the yield stress of the material before using the cylinder block are expected to exist.

If the copper bushing is not inserted, there is a high possibility that the zone of occurrence of the seizure and the operating area of the cylinder block overlap (**Figure 2**). As a result, if a failure occurs due to the seizure, the total non-operating time of the equipment becomes longer. This leads to a decline in equipment reliability. Therefore, if the durability of the cylinder block is not adversely affected even if the cost is somewhat high, it is considered necessary to insert the copper bushing even if residual stress occurs to improve the reliability of the main pump.

3. Methods for durability assessment of cylinder block

In general, standard, measurement, experiment and simulation can be considered as methods for durability assessment. The above methods will be described with respect to the cylinder block in which the residual stress is present.

In the case of the standard, the reliability of the evaluation result can be secured based on the authority of the standard, but the accuracy of the product with complicated shape is degraded in order to guarantee the performance. The reason is that the standard contains some assumptions and simplifications in the formulas for general use.

In case of measurement, there is an advantage that it can confirm the accurate state in real time when using, but there is a disadvantage that it is only possible to check the local state instead of the total state. In addition, there are many difficulties in measuring a cylinder block having a relatively small size as well as a rotating body. Also, there is a risk that the location of the high stressed portion, which is an evidence for durability safety, should be selected only by experience.

Experiments provide a relatively complete assessment of the product as a whole, but it takes time and expense to derive the results of the experiment. Especially, the improvement process and the short processing period, which are often derived when the product development stage is the basic design, make the experiment only used in the final product development stage. These characteristics make the experimental method difficult to use for the purpose of this study (predicting the residual stress and evaluating the durability thereof), which is the working area of the basic design.

Finally, in the case of simulation, the overall evaluation of the product is possible and it is possible to respond quickly to the improvement. However, as shown **Figure 3**, unlike the measurement and the experiment, simulation is difficult to consider stochastic effect. Stochastic effects of products not confirmed by

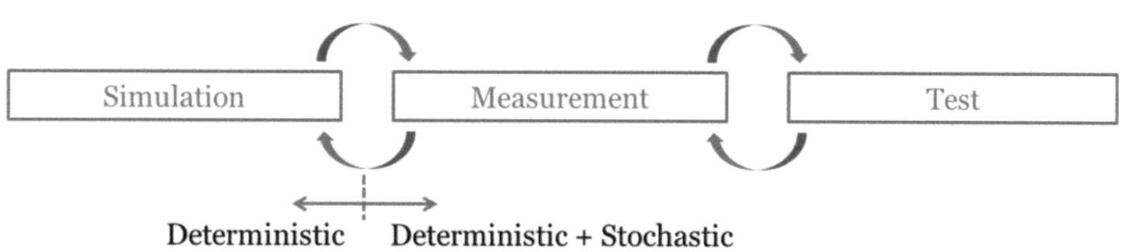

Figure 3.
Deterministic and stochastic in evaluation method.

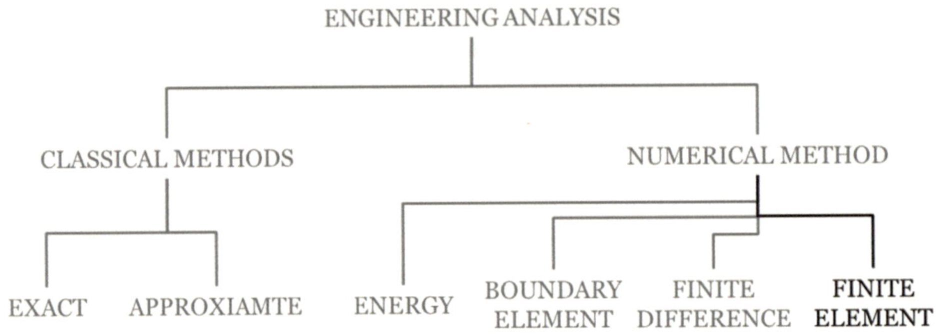

Figure 4.
Methods for engineering analysis.

simulation methods are expected to be confirmed in the final tests performed prior to final product launch.

The simulation used in this study is structural simulation. The Finite Element Method (FEM) was used as shown in **Figure 4**). The finite element method is one of the powerful methods to numerically solve wide engineering problems. Solve the equation of the virtual works (Eq. (4)) using the equilibrium equations and boundary conditions equation (Eqs. (1)–(3)) in the engineering problem domain. Simulations use the weighted residual method as a way to solve this equation of virtual work linearly. By using this weighted residual method, it gives an approximate solution of the matrixed problem. At this time, since the method used to solve the matrixed problem is the node and the element, the know-how of element setting that does not distort the coordinate system of problem is important for solving the problem.

3.1 Residual stress according to rates of shrink fit

As mentioned above, when the copper bushing is inserted into the cylinder block to prevent the sticking, residual stress is expected before use of the product. The amount of residual stress is expected to be proportional to the rate of shrink fit. The small rate of shrink fit is likely to cause detachment of the bushing, and the large rate of shrink fit is expected to adversely affect the durability of the copper bushing and cylinder block (**Figures 5** and **6**). Therefore, it is important to select the rate of

Figure 5.
Bushing expected departure direction [3].

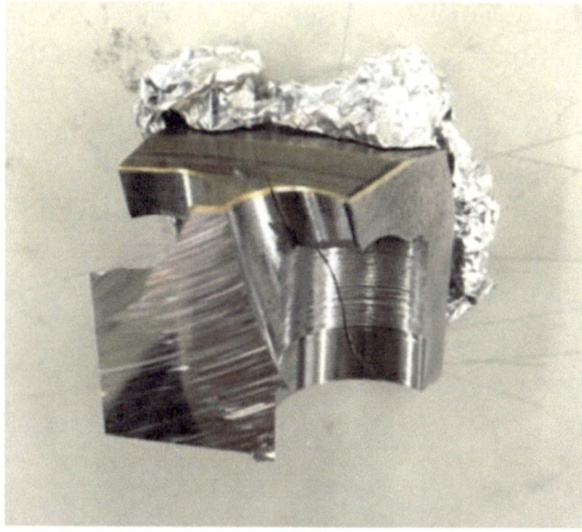

Figure 6.
Crack of cylinder block.

shrink fit suitable for the size and material of the cylinder block and copper bushing before use.

3.2 Material property check

The shape and material have a great influence on the residual stress as well as the rate of shrink fit when the copper bushing is pressed into the cylinder block. In order to predict the accurate residual stress, a characteristic test was performed on the material except for the shape selected through the design [4]. The reason for this is that, in the case of metallic materials, the material properties such as yield stress and tensile stress vary depending on the production method of the material. Therefore, in order to perform accurate durability assessment, the above-mentioned yield stress and tensile stress value should be basically confirmed. The test results of yield stress and tensile stress are shown in **Figure 7** and **Table 1**. Experimental results were derived from other common aspects that copper

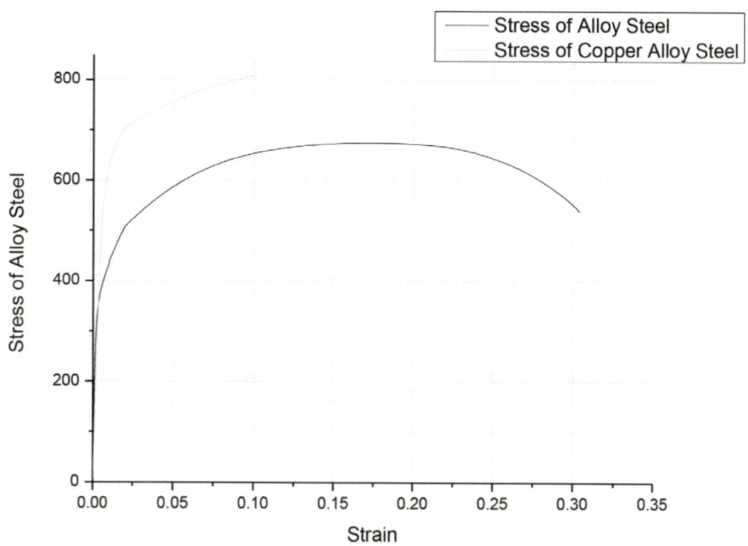

Figure 7.
Material property test results.

bushings generally exhibit the characteristics of ductile materials and alloy steels exhibit characteristics of brittle materials [5]. The results of the material properties of **Table 1** were used to evaluate the durability of the precise cylinder block. **Figure 7** shows the result of the measurement, not the literature, and the confirmed values are shown in **Table 1**.

The residual stress acts as a mean stress at the time of durability assessment, and safety is different as shown in **Figures 8** and **9**. Therefore, in **Figure 9**, the result of **Table 1** will be used to draw the Haigh diagram baseline clearly.

$$\sigma_a = \frac{\sigma_{max} - \sigma_{min}}{2} \text{(alternating stress)} \tag{1}$$

$$\sigma_m = \left(\frac{\sigma_{max} + \sigma_{min}}{2} \text{mean stress}\right) \tag{2}$$

In addition to the Haigh diagram, there are five other diagrams to easily assess the durability of a product and to illustrate it. The expressions for expressing diagrams are the same as **Table 2**. Goodman, Gerber, SAE, Soderberg and Modified Goodman shown in **Table 2** are shown in **Figure 10**. As you can see in **Figure 10**, Soderberg, Modified Goodman, SAE, Goodman and Gerber are listed in the most conservative order of product evaluation for durability. The most similar form to Haigh diagram is Modified Goodman.

The Haigh diagram (**Figure 11**) can be accurately draw as following:

Point 1: The right limit of the Haigh diagram is generally given by the tensile strength of the material.

Points 2 and 3: For ductile materials, the second point is defined as the intersection between the straight line $\sigma_o = \sigma_y$ and the straight line through the alternating stress limit of the material ($\sigma_{A,tsc}$, R = −1) and its pulsating stress limit ($\sigma_{A,tsc}$, R = 0).

	Alloy steel	Copper alloy steel
Tensile strength	194	226
Yield strength	100	161
Elongation (%)	25	5.45

Table 1.
Value of test results.

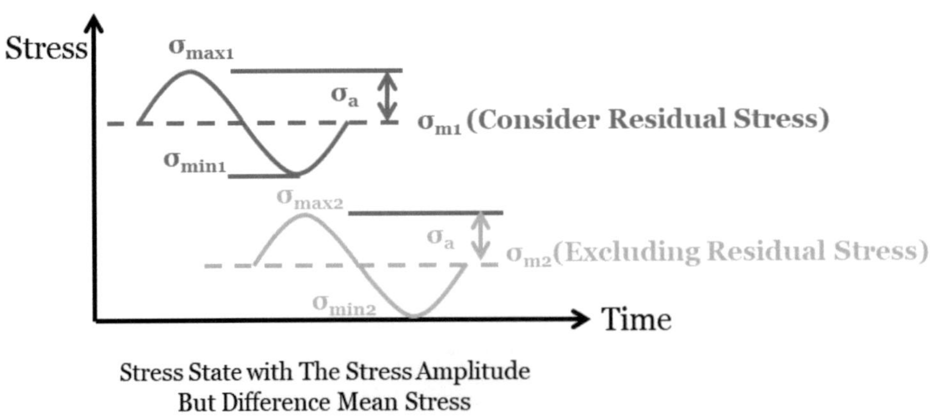

Figure 8.
Important of residual stress.

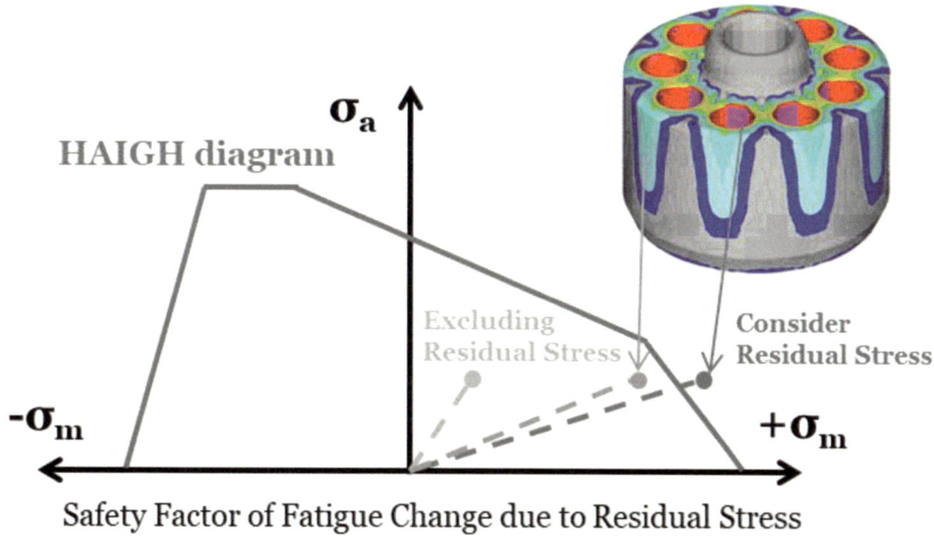

Figure 9.
Important of material property.

Point 3 is identical with point 2. For GG the points 2 and 3 are, as a result of experiments, defined as:

Point 2: $\sigma_m = 0.88 \cdot \sigma_{UTS}$, $\sigma_0 = 0.34 \cdot \sigma_{A,tsc}$ and.

Point 3: $\sigma_m = 0.76 \cdot \sigma_{UTS}$, $\sigma_a = 0.48 \cdot \sigma_{A,tsc}$.

Point 4: Pulsating stress limit (amplitude) of the material.

Point 5: Alternating stress limit of the material under tension/compression.

Point 6: For ductile materials, point 6 is defined as the intersection of the straight line $\sigma_1 = -\sigma_y$ with the lengthening of the straight line from point 4 to point 5. For GG an average inclination of the straight line of 30 degree is derived from known compression pulsating stress limits, which, together with a straight line $R = -\infty$, give point 6. If the compression pulsating stress limit of the material is known, point 6 can be derived from it.

Point 7: For GG, point 7 is determined by half of the length of an orthogonal straight line through the intersection of the compression-fracture border line with the straight line $R = -\infty$. For all other materials, point 7 is identical with point 6.

Point 8: It is the intersection of a horizontal straight line through point 6 with the straight line $\sigma_l = \sigma_{c,C}$.

Fracture theory	Goodman	Gerber	SAE
Equation	$\frac{\sigma_a}{S_e} + \frac{\sigma_m}{S_u} = 1$	$\frac{\sigma_a}{S_e} + \left(\frac{\sigma_m}{S_u}\right)^2 = 1$	$\frac{\sigma_a}{s_e} + \frac{\sigma_m}{S_f} = 1$
Yield theory	**Soderberg**	**Modified Goodman**	
Equation	$\frac{\sigma_a}{S_e} + \frac{\sigma_m}{S_y} = 1$	$\frac{\sigma_a}{S_e} + \frac{\sigma_m}{S_u} = 1 \left(for\ \frac{\sigma_a}{\sigma_m} \geq \beta\right)$ $\frac{\sigma_a + \sigma_m}{S_y} = 1 \left(for\ \frac{\sigma_a}{\sigma_m} \leq \beta\right)$	

Subscript notation: a = alternating, y = static tensile yield, m = mean, u = static tensile ultimate, e = modified material constant, $\beta = S_e(S_u - S_y)/S_u(S_y - S_e)$, σ_a = alternating stress in enviroment condition, σ_m = mean stress in enviroment condition, S_e = modified stress of material, S_f = fatigue limit of material in environment condition, S_y = yield stress of material, S_u = ultimate stress of material.

Table 2.
Types of HAIGH Diagram.

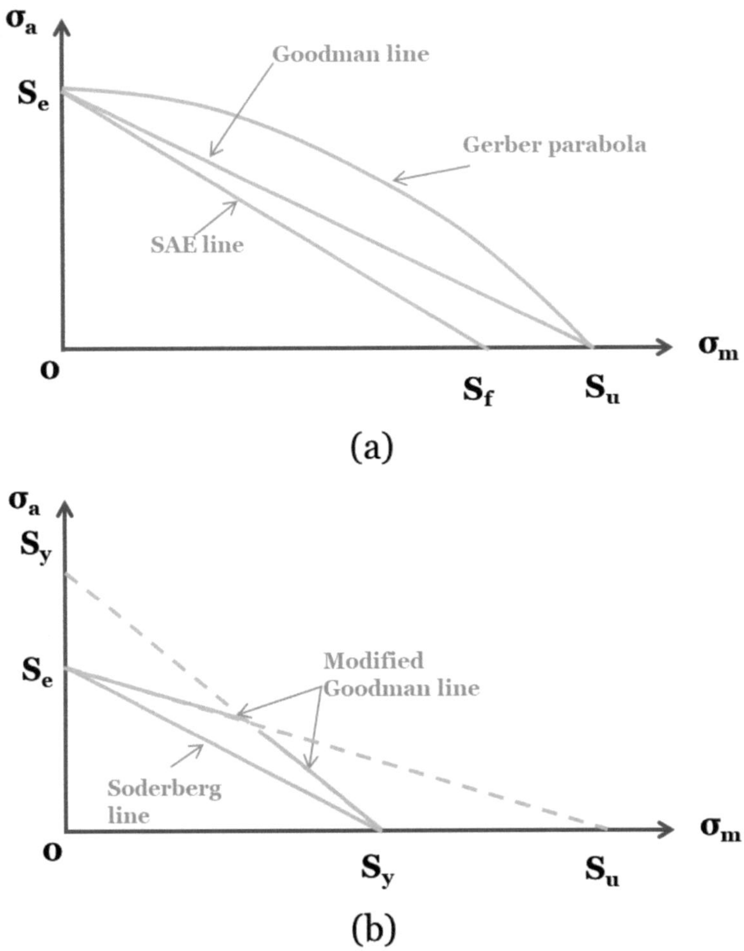

Figure 10.
Fatigue lines to various fracture theories. (a) Fracture theory, and (b) Yield theory.

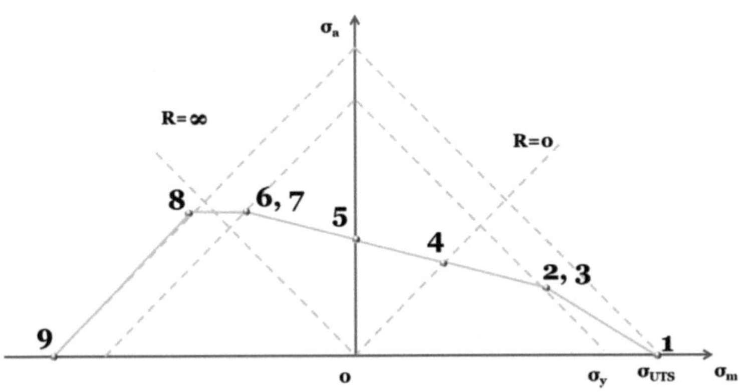

Figure 11.
Drawing method of Haigh diagram.

Point 9: The left limit of the Haigh diagram is given by the ultimate stress limit under compression of the material.

However, the S-N curve and the S-S curve can be changed by the size effect, the relative stress gradient, and the temperature effect, which are exactly the values of

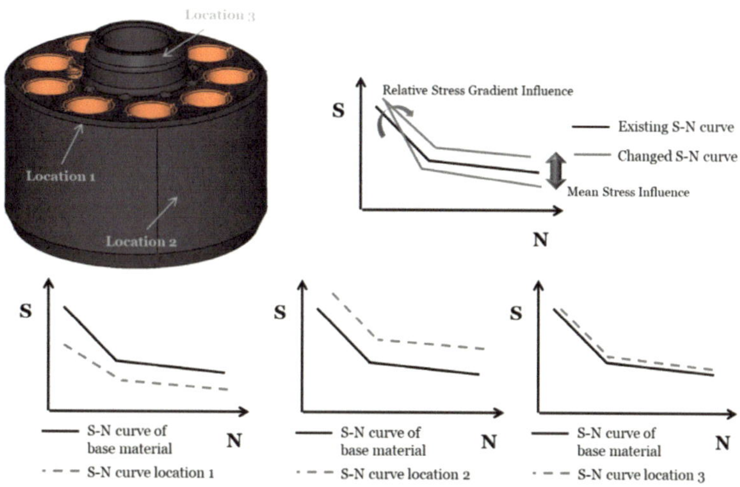

Figure 12.
Change curved because of fatigue parameter.

σ_a and Se necessary for drawing the Haigh diagram as shown in **Figure 12**. In order to consider the fatigue factor, more complicated calculations must be accompanied. Therefore, it is recommended to use the fatigue program for efficient and accurate fatigue safety factor evaluation.

3.3 Verification of simulation method

The purpose of this study is to quantitatively predict the residual stresses that are necessary before actual cylinder block for use. Therefore, it is important to set up an appropriate method of simulation, which is an economic evaluation method, before production. For this purpose, this study compares the surface pressure distribution of the simulation with the surface pressure prediction calculated as (**Figure 13** and Eq. 3) [6].

$$\mathrm{p} = \frac{\delta}{\frac{b}{E_h}\left(\frac{b^2+c^2}{c^2-b^2} + \nu_h\right) + \frac{b}{E_s}\left(\frac{a^2+b^2}{b^2-a^2} - \nu_s\right)} \; \textit{Face Pressure} \tag{3}$$

E_h = modulus of elasticity for hole material,
ν_h = Poisson's ratio for hole material,
E_s = modulus of elasticity for shrink material,
ν_s = Poisson's ratio for shrink material
which is obtained by setting the following stress conditions (**Figure 14** and Eqs. (4) and (5)).

$$\frac{d\sigma_r}{dr} + \frac{\sigma_r - \sigma_\theta}{r} = 0 \; \textit{Equilibrium Equation} \tag{4}$$

$$\varepsilon_r = \frac{du}{dr}, \varepsilon_\theta = \frac{u}{r} \; \textit{Strain} \tag{5}$$

σ_r = radial stress,
σ_θ = tangential stress,
$\boldsymbol{\varepsilon_r}$ = radial strain,
$\boldsymbol{\varepsilon_\theta}$ = tangential strain,
$\boldsymbol{u}$ = displacement of radial direction

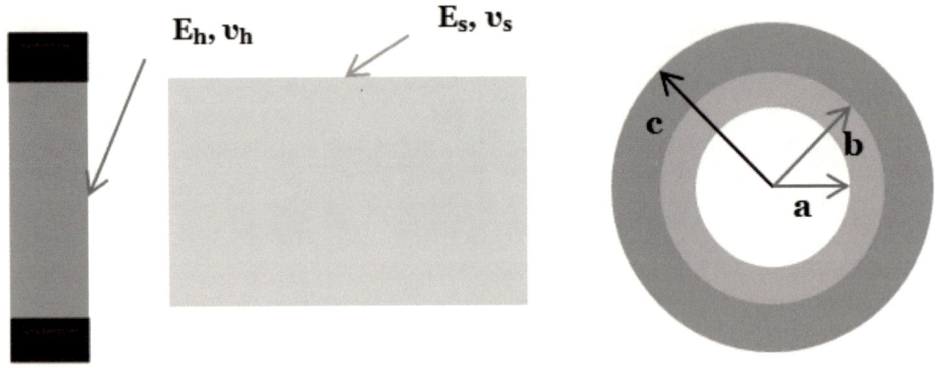

Figure 13.
State of shrink fit.

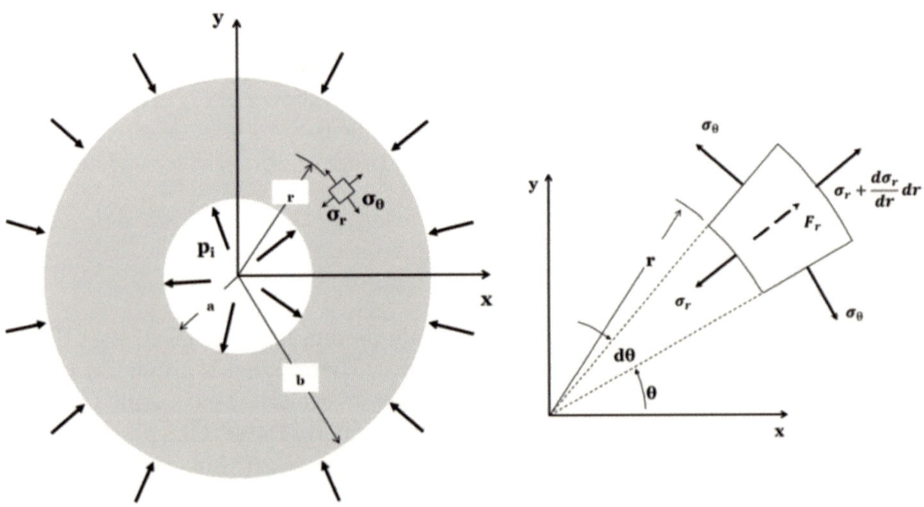

Figure 14.
Stress state.

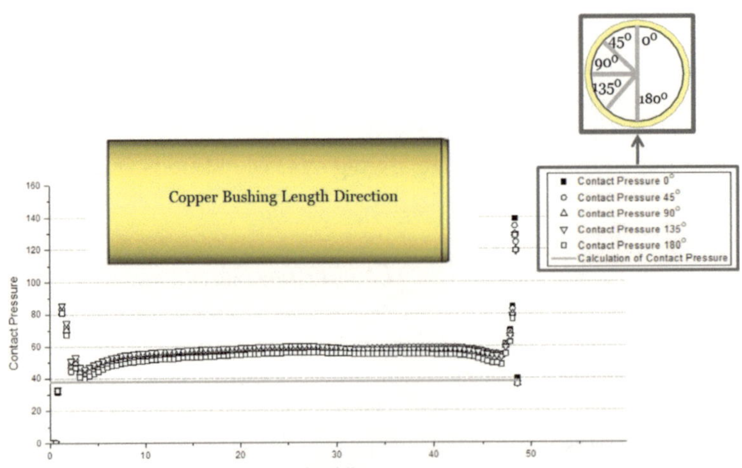

Figure 15.
Check of surface pressure.

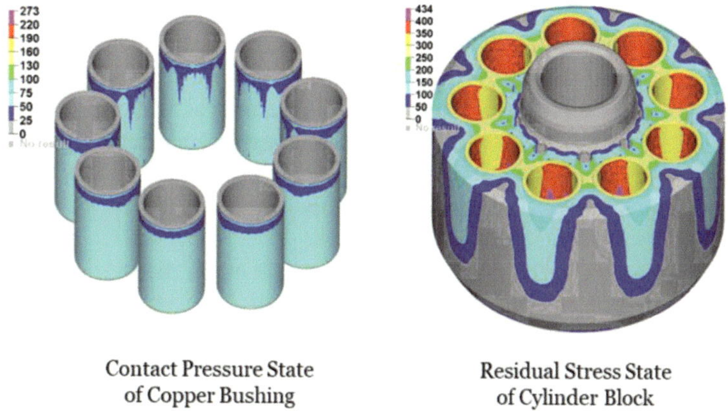

Figure 16.
State of contact pressure and residual stress.

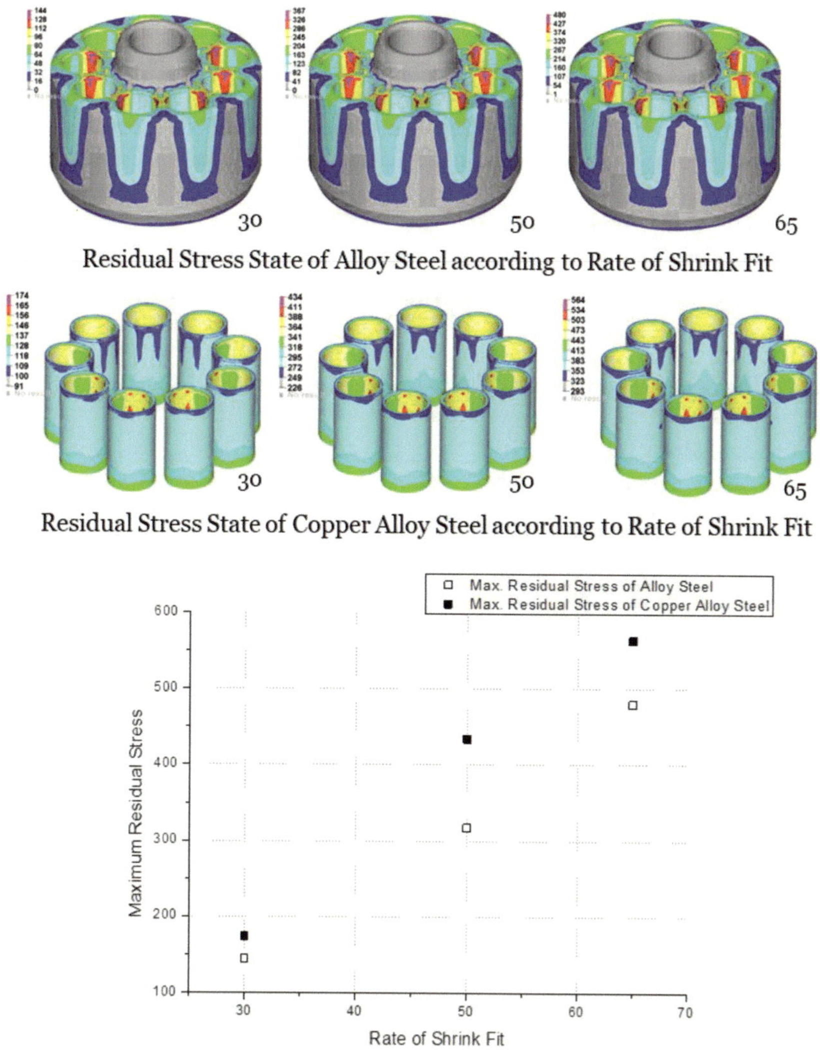

Figure 17.
Relationship between rates of shrink fit and residual stress.

The resulting equation assumes axial symmetry and does not take into account shear stress inherently. In addition, the constant surface pressure in the longitudinal direction is calculated, so it is not suitable for the shape of the cylinder block which is the object of this study, but it is judged to be suitable as the relative standard of the surface pressure to be derived from the structure simulation.

3.4 Estimate residual stress using simulation

The residual stress of the derived cylinder block using the structural simulation is shown in **Figure 15**. In order to determine the suitability of the simulation results, the surface pressure of the bushing was confirmed by angle. The surface pressure distribution of the simulation results follows the general shape [7] of the Hertz contact theory and most of the length of the surface pressure is higher than the constant value derived from the numerical formula (Eq. 5). The reason is that the shape of the cylinder block is not perfectly symmetrical and shear stress is generated almost everywhere (**Figures 16** and **23**).

In addition, the shape and numerical value of the residual stresses were checked within the range of possible shrink fit to confirm the aspect of the change in the residual stress with respect to the rate of shrink fit [8]. As a result, as shown in **Figure 17**, the shape of the residual stress was less dependent on the rate of shrink fit, and the numerical value of the residual stress was almost linearly related to the rate of shrink fit. Therefore, it was possible to set the rate of shrink fit and the residual stress in a linear relationship within the range of the producible shrink fit. In this study, it was possible to select the rate of shrink fit optimized for the use environment by using this relationship. If the relationship between the rate of shrink fit and the residual stress was nonlinear, then the simulation was additionally performed to find out the sections having linear relationship to find the optimal design.

4. Durability assessment considering residual stress

The residual stress results derived from the above simulation are high as the yield stress of the cylinder block and the bushing, so the residual stress was used as the mean stress before the cylinder block was used for accurate durability assessment. Unlike the above situation, when the residual stress is low, there are few reasons to consider it. This is because, as shown in **Figure 18**, there is no difference in the durability assessment result according to the load size when the residual stress is low. In conclusion, the cylinder block, which is the subject of this study, has

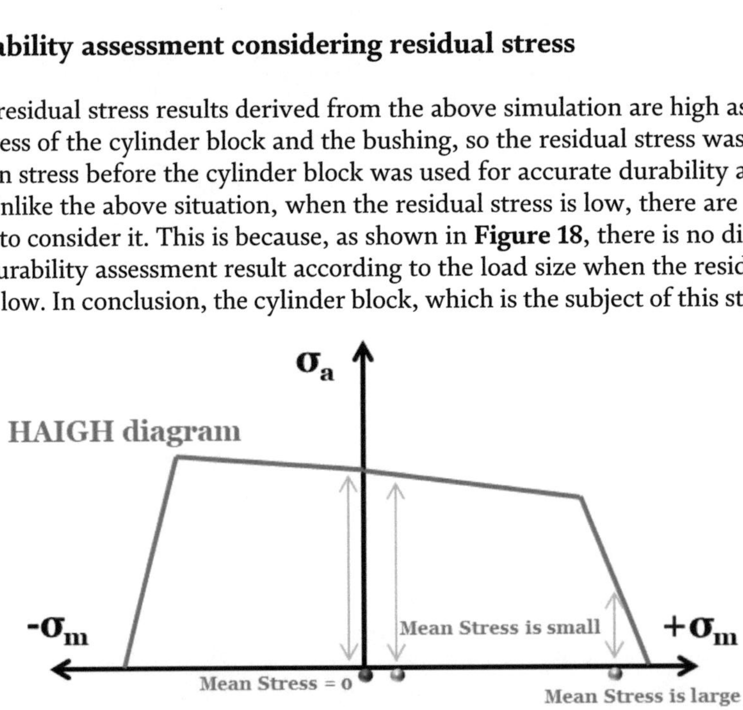

Figure 18.
Mean stress influence on HAIGH diagram.

a high residual stress value. Therefore, in this study, the durability assessment was performed considering the residual stress. In addition, in this section, realistic load and boundary conditions considering the use environment are examined to improve the reliability of the structural simulation results.

4.1 Simulation condition of cylinder block

The cylinder block is fixed from the shoe plate of the main pump and rotates about the rotation axis. At this time, the shoe plate has a certain angle, and the stroke of the piston is generated through this. As a result, during one revolution of the cylinder block, the working fluid flows into the piston through the top dead center and the bottom dead center, and the high-pressure operating fluid what the user wants is discharged [9]. At this time, the cylinder block can be divided into a low-pressure suction portion and a high-pressure discharge portion, which is a main working load condition acting on the cylinder block. In order to quantitatively confirm the magnitude of this high pressure, this study uses the swash plate angle of the main pump, the position of the piston end when the cylinder block is rotated and the AMESim program. As a result, realistic loads such as **Figures 19** and **20** were obtained.

In addition, the boundary condition as shown in **Figure 21** was set in consideration of the operating condition of the cylinder block. The boundary conditions were fixed in the axial direction where the stopper was located, and the angle of the spline portion was fixed to prevent rigid body motion of the cylinder block. In this case, the accuracy of the simulation is ensured by applying the centrifugal force to the influence of the rotational speed of the cylinder block while rotating the cylinder block.

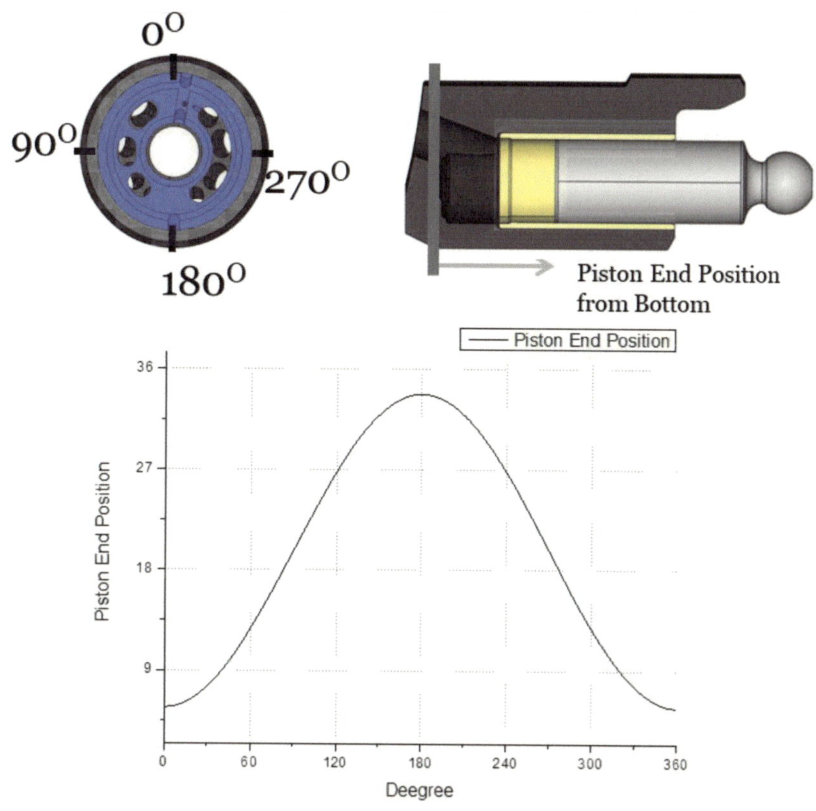

Figure 19.
Piston end position from bottom for simulation.

Figure 20.
Working load of main pump.

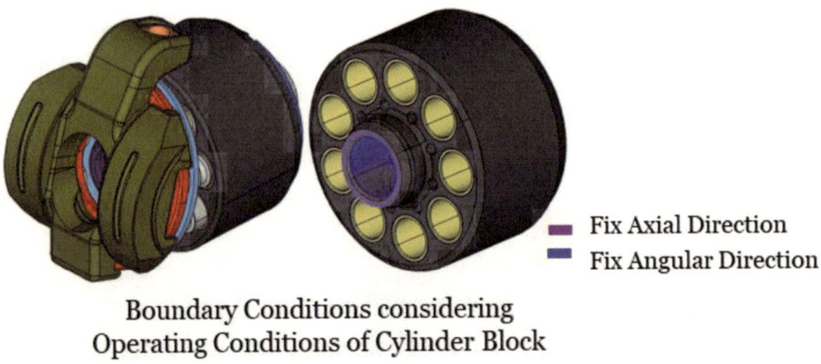

Figure 21.
Boundary condition for FEM.

4.2 Comparison of simulation results considering residual stress

During operation of the main pump, the load acting on the cylinder block does not act as a cyclic symmetry, and the load distribution during rotation varies. Therefore, in this study, the time when the cylinder block receives the highest load during rotation is selected and analyzed (**Figure 22**). The reason for this is that the cylinder block will rotate and receive a load at this peak time periodically at the lowest stress (0 MPa).

The results of the residual stress analysis obtained above are mapped to the structural simulation model and the results shown in **Figure 23** can be obtained by using the above described simulation conditions and methods. It can be confirmed that there is a considerable difference by comparing it with the stress value of the simulation method which does not consider the residual stress (**Figure 24**).

4.3 Durability assessment of cylinder block considering residual stress

The durability assessment of the cylinder block was finally evaluated by using the stress simulation result considering the residual stress, the use environment and the allowable stress which is one of the characteristic values of the material. At this

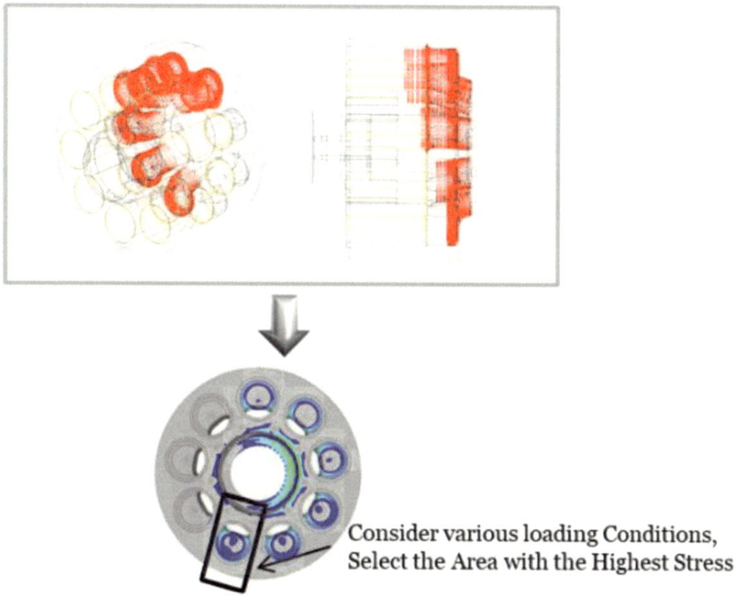

Figure 22.
Extreme situation during usage.

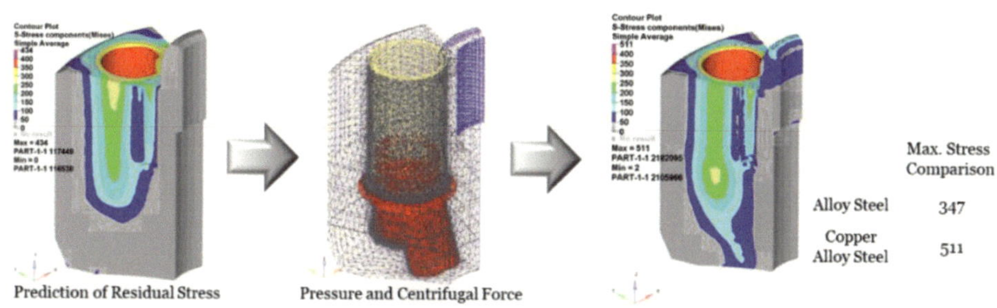

Figure 23.
Durability assessment results of cylinder block.

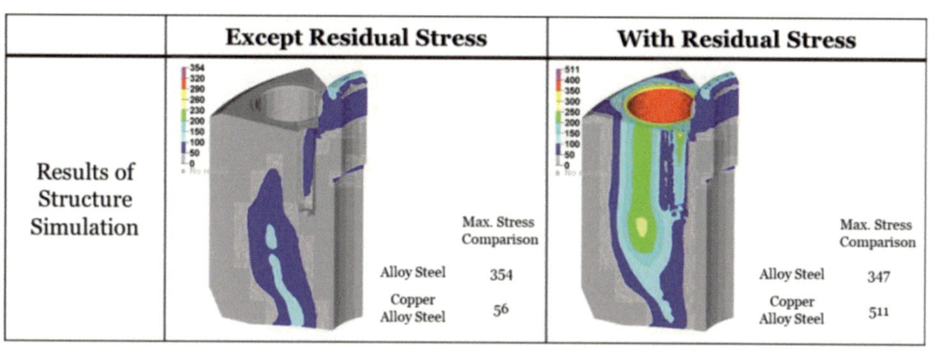

Figure 24.
Comparison of durability assessment results.

time, among the many durability assessment methods, the mean stress is constant in the cylinder block, and the stress condition during use is relatively simple, so that it is calculated according to the criterion of σ_m = C in **Figure 25**. Finally, in order to improve the accuracy of fatigue safety assessment, size factor, slope of surrounding stress, survival probability and surface roughness were considered [8]. As a result, quantitative fatigue safety factor as shown in **Figure 26** was obtained.

The residual stresses and assessment methods mentioned in the above studies can often be different from the residual stresses in welding and their more complicated assessment methods. This is because, in the case of welding, the change of the metal structure due to the introduction of heat energy occurs, and thus a complicated residual stress is formed. However, the residual stress generated in this study is not related to the residual stress due to the temperature change and is related to

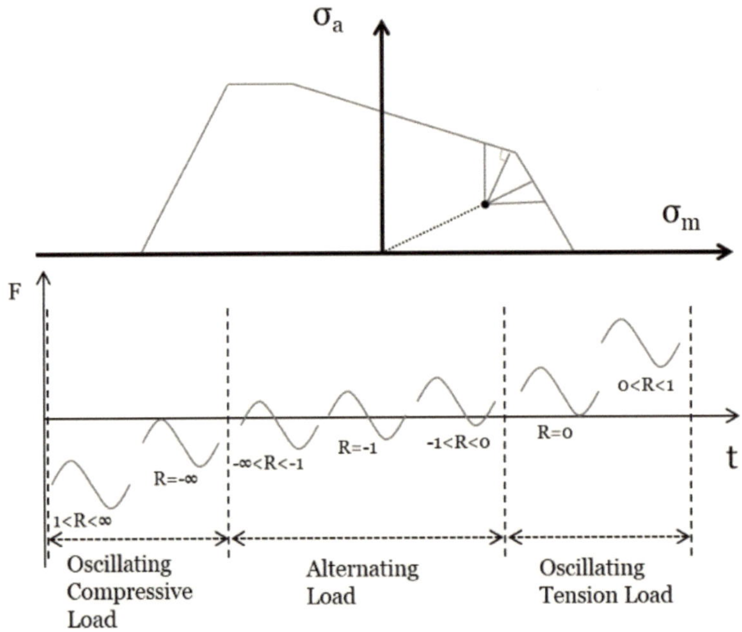

Figure 25.
Method for durability assessment.

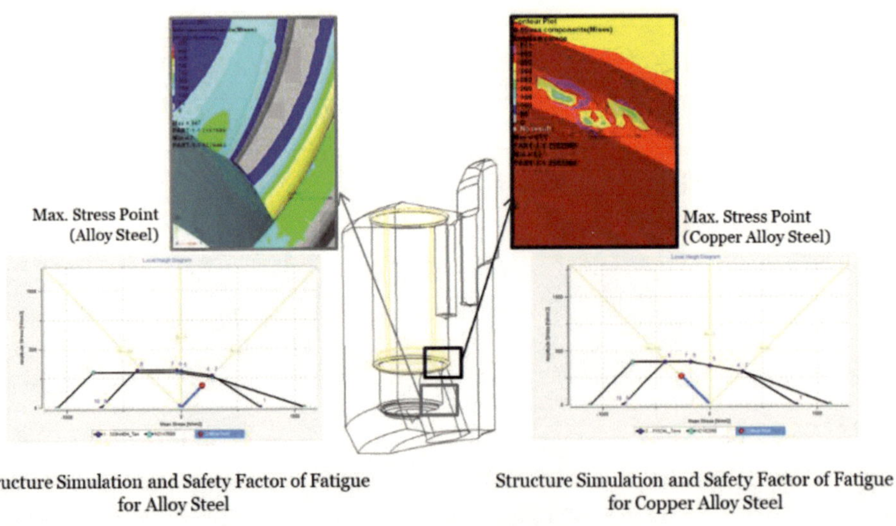

Figure 26.
Quantitative result values.

the shrinkage and relaxation of the existing material. Therefore, it is possible to perform a definite durability assessment only by the basic relationship between the basic load and the material (**Figures 27** and **28**) [10].

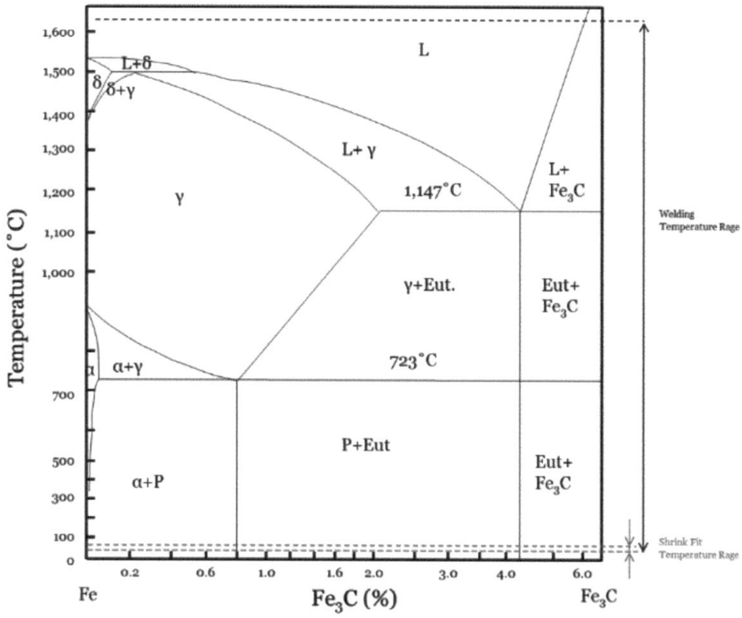

Figure 27.
Phase diagram of Fe-Fe_3C.

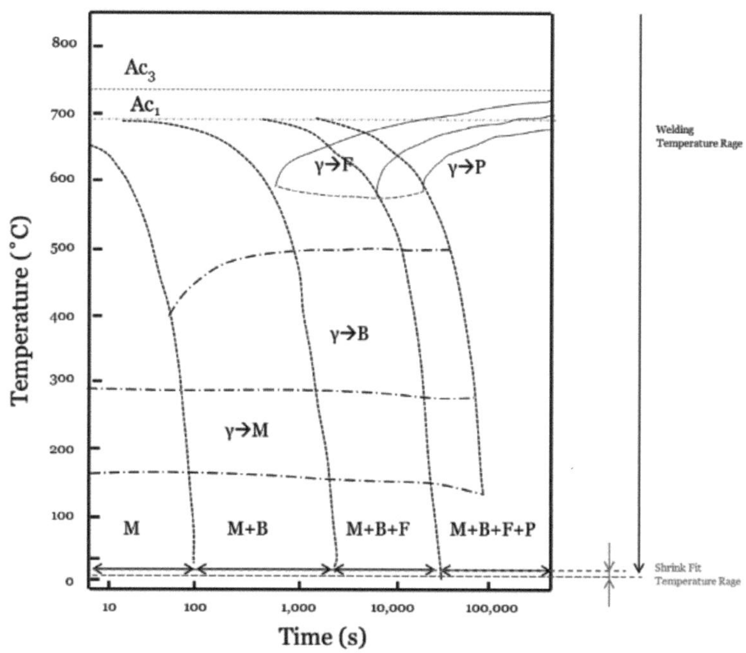

Figure 28.
Continuous cooling transformation diagram.

5. Conclusion

This study was carried out to find methods for quantitatively confirming the durability of the product during the increasingly diverse production process. And, the cylinder block in which the bushing was inserted for seizure was selected as the standard product. As a result, it was possible to establish durability assessment methods and processes for machine parts where large residual stress is generated by shrink fit during production process. The results of this study will contribute to the durability assessment considering the residual stress of machine parts which does not change the material structure due to thermal energy input unlike welding.

In addition, this study examined existing durability assessment methods and confirmed the merits and demerits of evaluation methods. Based on these advantages and disadvantages, this study also presented limitations of the durability assessment method considering the residual stresses presented. This will prevent indiscreet use by specifying the use environment of the durability assessment considering residual stress as a result of this study. In addition, a verification method is proposed to use the simulation for complex shapes and to ensure the reliability of the simulation results. This will provide efficient development and forecasting directions for the industry unlike current development of shrink fit that has been developed depending on existing people and facilities.

Author details

Byung-Ik Kang
Korea Gas Safety Corporation (KGS), Bukbusunhwando-ro, Nam-gu, Ulsan, South Korea

*Address all correspondence to: byoungik5@gmail.com

References

[1] Jung DY, Koe CH. Tribology. Korea: GS Intervision; 2014. pp. 181-182

[2] Peterson RE. Stress Concentration Factors. Korea: John Willey & Sons; 1974. pp. 15-19

[3] Kang BI. Durability assessment of cylinder block based on residual stress. Transactions of the Korean Society of Mechanical Engineers, A. 2018;**42**: 605-610

[4] KRAUSS. Principles of Heat Treatment of Steel. United States of America: ICT; 1993. pp. 1-142

[5] Human Resources Development Service of Korea. Metallic Material. Korea: Human Resources Development Service of Korea; 2005. pp. 135-185

[6] Ugural AC. Mechanical Design. Korea: McGrawHill; 2005. pp. 630-649

[7] Barber JR. Elasticity. United States of America: Kluwer Academic Publishers; 1991. pp. 120-220

[8] Adams V, Askenazi A. Finite Element Analysis. United States of America: Cengage; 2010. pp. 347-379

[9] Jung DS, Kim DS, Lee YB. Reliability Evaluation Technique for the Main Pump of a Excavator. Vol. 4. Korea: KSFC; 2012. pp. 88-96

[10] The Korean Welding and Joining Society. Handbook of Welding Joint I. Korea: The Korean Welding and Joining Society; 2017